CANALS — A NEW LOOK

CANALS
A New Look

Studies in honour of
CHARLES HADFIELD

Edited by
Mark Baldwin
and
Anthony Burton

Phillimore

1984

Published by
PHILLIMORE & CO. LTD.
Shopwyke Hall, Chichester, Sussex

ISBN 0 85033 516 7

Typeset in the United Kingdom by:
Fidelity Processes - Selsey - Sussex

Printed and bound in Great Britain by
BILLING & SONS LTD
Worcester, England

CONTENTS

LIST OF PLATES
(between pages 22 and 23)

LIST OF TEXT ILLUSTRATIONS

ACKNOWLEDGEMENTS

Photographs for the plate section were provided by the following: Harry Arnold: No. 31; B. R. Berkeley: No. 1; Bodleian Library, Oxford: No. 4; British Library: No. 2; British Waterways Board: No. 8; Richard Bryant; Nos. 14, 16, 19, 20, 23, 24, 26; Leslie Bryce: Nos. 15, 18, 22, 28, 29; Cambridgeshire Record Office: No. 5; Gérard Presse: No. 10; David Grindley: No. 30; Oscar & Peter Johnson Ltd., Lowndes Lodge Gallery, 27 Lowndes St., London (photo by Joyce Gurr): No. 3; KLM Aerocarto n.v.: No. 9; Port of Rotterdam: No. 6; Derek Pratt: Nos. 12, 13, 21, 25; Russell Ray Studios: No. 7; Shell UK Ltd: Nos. 32, 33; Dr. Roger Squires: Nos. 34, 35; Peter White: Nos. 11, 17, 27.

The illustrations for the text were provided as follows: the British Waterways Board gave permission to use Figs. 5.1, 5.2 and 5.3; David Edwards-May drew the diagrams for Chapter 3; Anne Langford drew those for Chapters 1 and 2; Richard C. Packer took photographs for the line illustrations for Chapters 4 and 7 (all originals from the collection of Dr. and Mrs. Mark Baldwin); and Peter White drew 5.4 and 5.5 (as well as 5.1).

The photograph on the front cover was taken by Joyce Gurr, with the kind permission of Oscar & Peter Johnson Ltd. of the Lowndes Lodge Gallery, 27 Lowndes St., London.

We are most grateful to all the individuals and bodies named above for their help.

CHARLES HADFIELD:

A BIOGRAPHICAL INTRODUCTION

MOST READERS of this book will know Charles Hadfield as a writer on canals — indeed it is our present purpose to honour him for his very substantial contribution in this field. However, as the list of his published works shows, his literary achievements are by no means limited to inland waterways. This brief biographical sketch therefore has twin aims: to outline Charles' extensive involvement with waterways (an involvement going far beyond writing about them) and to explain how some of his other books and articles came to be written. I have, of course, been dependent on Charles for most of the relevant information, largely through the medium of his unpublished autobiography, kindly lent for this purpose.

Ellis Charles Raymond Hadfield was born in 1909 in South Africa and, although his schooling was started there, it was completed in England at Blundell's School, Tiverton. In 1928 he entered St Edmund Hall, Oxford, with an Exhibition in English. A change of course led to his obtaining a degree in Economics, a subject not immediately suggestive of a particular career on graduation. After several uncertain starts, Charles joined 'Stonewall' Jackson, a small-time secondhand bookseller who proved more colourful than competent. The venture was not what a bank manager would term a success, but it left an indelible mark on Charles — the 'Sign of the Book'. After four years the partnership was dissolved and Charles joined the London staff of the Oxford University Press; it was here that he met Alice Mary Miller, also to be an author, whom he later happily married. In 1939 he took charge of the Juvenile Department of OUP, his task of modernisation being interrupted by war service as an auxiliary fireman in the River Service of the London Fire Brigade.

While at Oxford, Charles had developed a deep interest in politics and this led to his standing as a Labour Candidate in the Borough Council Elections of 1934, and being elected for the Harrow Road Ward of Paddington. At 25, he was one of London's youngest Councillors but soon became Whip under the leadership of James MacColl. He remained on the Council throughout his bookshop and Fire Service days, and not surprisingly became involved in the Fire Brigades Union in 1940, eventually making history when the paper he started, the *River Service Bulletin*, became the country's first printed local union journal.

In 1942, he successfully applied for a transfer to the Home Office Fire Staff to work on the seven-volume *Manual of Firemanship*. This was written by a staff of three (the others being Frank Eyre and V. J. Wilmoth) and the worth of their work may be judged from the fact that, albeit much revised, it is still today the basic text on firefighting.

Leaving OUP in 1946, Charles moved to the Central Office of Information, to become its Director of Publications. With a large staff and broad responsibilities,

1

this position involved the production of numerous booklets and magazines in a variety of languages, many for distribution within newly liberated Europe and Asia. Amongst the historic tasks it threw up was the preparation of two explanatory pamphlets on the Labour Government's revolutionary schemes — National Insurance and the National Health Service. Later promotion to Controller (Overseas) increased the range and magnitude of his responsibilities, which included master-minding the arrangements for the British Government pavilion at the 1958 Brussels International Exhibition, a success for which he was awarded the CMG in 1959. In 1962, Charles left the Civil Service, an early retirement to allow time to nurture the two-year-old publishing business of David & Charles, and to persevere with his canal research.

Charles' interest in canals dates from his Devon schooldays, when walking his dog along the towpath of the Grand Western Canal had stimulated a mild curiosity about its history. By chance, the family solicitor had been able to satisfy this curiosity by producing from his office a box of Grand Western papers. From this casual beginning sprang a lifelong interest in canals. Amongst its early products were Charles' first published work on canals in 1942, and his first canal book (with Frank Eyre) in 1945. In 1944, L. T. C. Rolt's *Narrow Boat* was published; on the strength of a shared interest, Charles wrote to Rolt and this led eventually to a meeting in 1946 in Robert Aickman's flat, at which the Inland Waterways Association was founded. Aickman was elected Chairman, Hadfield Vice-Chairman, Rolt Secretary, and Eyre Treasurer. As is often the case, when the IWA grew from being a microscopic to a small body, its members spent more time fighting face to face than shoulder to shoulder. In 1951, a tragic conclusion was reached — many members were expelled, including both Tom Rolt and Charles Hadfield. Undeterred, in 1954 Charles helped to form the Railway and Canal Historical Society, which continues to flourish.

In 1960, many pieces of Charles' jigsaw suddenly interlocked when the publishing firm of David & Charles was formed. David St John Thomas was then 30 years old, a farmer, a freelance journalist and broadcaster, and an expert on railways. Charles contributed his accumulated experience of almost every facet of publishing, and his expertise on waterways. From cautious beginnings, when the two partners backed their knowledge of their own subjects, the firm has grown explosively to become one of Britain's biggest non-fiction publishers. Some regret the passing of the early specialist days; I well remember, 20 years ago, reading with pleasure the advertisements that D&C used to place at the bottom of the front page of the *Guardian* telling of their latest ideas. The facsimiles were perhaps the most interesting: not just obvious classics like Priestley's *Navigable Canals*, but exciting eccentricities like telephone directories, shop catalogues and railway timetables. Charles remains an editor and author for D&C, although he sold his partnership in 1964.

The year 1962, in which Charles retired from the Civil Service, also saw the establishment of the British Waterways Board on which Charles served from its very first meeting in 1963 to the end of 1966. Initially at least, Hadfield and Admiral Parham were the only Board members who knew anything about canals, and so were in a position to provide a lead during the Board's early years when significant policy decisions were evolving. Charles actively influenced many of these; his hand can be seen in *The Future of the Waterways* (1964), an important stage in formulating the policy which eventually led to the classification in the 1968 Transport Act of the nationalised waterways as Commercial, Amenity or Remainder. Another of his suggestions, and one whose very success bred its own demise, was the introduction of a barge-carrying system appropriate to the waterways of the north-east Midlands and Humberside. BACAT came, worked, but was conquered by the Hull dockers, whose campaign against BACAT led to its withdrawal in 1975. The improvement scheme for the Sheffield & South Yorkshire Navigation, eventually started in 1979, stems directly from a paper he presented to the Board in the early 1960s (although, of course, pleas for improvement date back to the last century).

In 1971, he was invited to rejoin the IWA by a later generation who knew little and cared less about the earlier schisms. He accepted eagerly, and soon had IWA Council backing for a small committee (later known as the Inland Shipping Group) to promote the ideas of modern commercial waterways. Appreciating that such ideas also needed support from the trade, he suggested first to the Inland Shipping Group, and then to BWB, that the latter should organise a waterways conference. The idea was accepted, and led to Freightwaves 75, at which the National Waterways Transport Association was launched. Charles and Harry Grafton of BWB between them did most of the preliminary organising work for NWTA, Charles chairing the steering committee established for the purpose.

In the last 20 years, therefore, Charles Hadfield has been personally responsible for, or associated with, many important waterway developments. In addition, he has become the country's leading canal historian. His *Canals of the British Isles* series, conceived in 1955, was completed in 1977 when the last of its 13 titles was published. Charles wrote seven, collaborated on two, and edited the other four. He also launched D&C's *Inland Waterways Histories* series, editing several of the titles. A glance at the list of his published work shows (to date) 25 canal books (many of which have been reprinted) and 33 shorter publications. Currently, he is working on another book which will also meet an evident need – a world history of canals.

Despite this range, it is for the *Canals of the British Isles* we should be truly grateful. This series, an essential reference for any transport or economic historian, records in detail the basic histories of virtually all of the hundreds of navigations of the British Isles. Almost invariably it has been primary sources which have provided the raw material: minute books, ledgers, letter books, account books, deposited plans, contemporary newspapers and so on. By using these, and by meticulously recording his use, Charles Hadfield has provided a reliable and thorough history of an important element of our heritage. Many have built and will continue to build upon his work, confident in the soundness of the foundation he has so painstakingly laid. Future historians will doubtless bless the man who tackled – and completed – such a daunting task.

MARK BALDWIN

Imperial College, October 1983

BIBLIOGRAPHY OF CHARLES HADFIELD'S PUBLISHED WORK
(to October 1983)

1 Waterways

1.1 Books and Booklets

1 EYRE, Frank, & ——
English rivers and canals. London: Collins, 1945. (Britain in Pictures Series No. 84)
2nd impression 1947.
Hadfield's first book on canals. Despite the impression given on the dustwrapper, Hadfield did write part of the text.

2 *British canals: an illustrated history*. London: Phoenix House, 1950. Republished by Readers' Union 1952. 2nd edn. 1959; 2nd impression 1962. 3rd edn. 1966, incorrectly described on title page as 3rd impression of 2nd edn. 3rd and later editions published by David & Charles, Newton Abbot. 4th edn. 1969; 2nd impression 1972. 5th edn. 1974. 6th edn. 1979. 5th and 6th edns. include Ireland. 4th, 5th and 6th edns. also published in paperback. 2nd and 4th edns. republished by A. M. Kelley (New York) in 1968 and 1969.

3 *Introducing canals: a guide to British waterways today*. London: Benn, 1955.

4 *The canals of Southern England*. London: Phoenix House, 1955. Later expanded to form Items 11 and 14.

5 *The canals of South Wales and the Border*. Cardiff: University of Wales Press; London: Phoenix House, 1960. 2nd impression 1960. 2nd edn. published by David & Charles (Newton Abbot) 1967, in conjunction with University of Wales Press; 2nd impression 1977.

6 —— & NORRIS, John
Waterways to Stratford. Dawlish: David & Charles; London: Phoenix House, 1962.
2nd edn. 1968.

7 *Canals of the world*. Oxford: Blackwell, 1964.

8 *Canals and waterways*. Newton Abbot: Raleigh Press, 1966. (Raleigh Press Brief Guide No. 22).

9 *The canals of the East Midlands (including part of London)*. Newton Abbot: David & Charles, 1966. 2nd edn. 1970.

10 *The canals of the West Midlands*. Newton Abbot: David & Charles, 1966. 2nd edn. 1969. 2nd edn. republished by A. M. Kelley (New York) 1969.

11 *The canals of South West England*. Newton Abbot: David & Charles, 1967. Republished by A. M. Kelley (New York) 1968.

12 *The canal age*. Newton Abbot: David & Charles, 1968. 2nd impression 1969. Paperback edn. published by Pan Books, 1971. 1st edn. republished by F. A. Praeger (New York) 1969. Readers' Union edn. 1969. 2nd edn. 1981.

13 —— & STREAT, Michael
Holiday cruising on inland waterways. Newton Abbot: David & Charles, 1968. 2nd edn. 1971, also published as paperback by Pan Books (London) 1972.

14 *The canals of South and South East England*. Newton Abbot: David & Charles, 1969. Republished by A. M. Kelley (New York) 1969.

15 —— & BIDDLE, Gordon
The canals of North West England (2 vols.). Newton Abbot: David & Charles, 1970.

16 *Canal enthusiasts' handbook 1970–71* (ed.). Newton Abbot: David & Charles, 1970.

17 *The canals of Yorkshire and North East England* (2 vols.). Newton Abbot: David & Charles, 1972–3.

18 *Canal enthusiasts' handbook No. 2* (ed.). Newton Abbot: David & Charles, 1973.

19 *Introducing inland waterways*. Newton Abbot: David & Charles, 1973. Paperback edition 1973.
20 DOERFLINGER, F., ──, et al.
   Barges or juggernauts? A national commercial waterways development projection. London: Inland Waterways Association, 1974. 2nd edn. 1974; 2nd impression 1975.
21 *Waterways sights to see*. Newton Abbot: David & Charles, 1976.
22 CLINKER, C. R., & ──
   The Ashby-de-la-Zouch Canal and its railways. Bristol: Avon–Anglia, 1978. Reprint of Item 35.
23 *Inland Waterways*. Newton Abbot: David & Charles, n.d. [1978].
24 ── & HADFIELD, Alice Mary
   Afloat in America. Newton Abbot; David & Charles, 1979.
25 ── & SKEMPTON, A. W.
   William Jessop, engineer. Newton Abbot: David & Charles, 1979.

1.2   *Papers, articles and other short contributions*

26 Canals between the English and the Bristol Channels. *Economic History Review*. 12 (1) and (2). 1942. 59–67.
27 The Thames Navigation and the canals, 1770–1830. *Economic History Review*. 14 (2). 1944. 172–9.
28 James Green as canal engineer. *Jnl. Transport Hist*. 1 (1). 1953. 44–56.
29 ── & CLINKER, C. R.
   The Ashby Canal: importance of its tramroads. *Modern Transport*. 7 August 1954. 5.
30 Report on visit to the Grand Junction Canal. *Jnl. Rly. Canal Hist. Soc*. 1 (2). April 1955. 8–9.
31 An approach to canal research. *Jnl. Rly. Canal Hist. Soc*. 1 (3). July 1955. 23–6.
32 Sources for the history of British canals. *Jnl. Transport Hist*. 2. 1955–56. 80–9.
33 Passenger boats on the Chester Canal 1775–1806. *Jnl. Rly. Canal Hist. Soc*. 2 (3). May 1956. 34–6.
34 The Cromford Canal. *Jnl. Rly. Canal Hist. Soc*. 3 (3). May 1957. 49–51.
35 CLINKER, C. R., & ──
   The Ashby-de-la-Zouch Canal and its railways. *Trans. Leics. Archaeol. Hist. Soc*. 34. 1958. 53–76. Later reprinted as a booklet (Item 22).
36 Canals: Inland waterways of the British Isles. Article in *A History of Technology*. Vol. IV. Oxford: Clarendon Press, 1958. 563–73.
37 Canals. Article in *Victoria County History of Wiltshire*. Vol. 4. London: Oxford University Press, 1959. 272–9.
38 The Grand Junction Canal. *Jnl. Transport Hist*. 4 (2). 1959. 96–112.
39 Writing railway and canal history. *Jnl. Rly. Canal Hist. Soc*. 8 (6). November 1962. 91–5.
40 Britische Kanal- und Fluss-schiffahrt. *Wasser und Boden*. Hamburg. March 1964. 83–7.
41 Water transport, Inland. Article in *Encyclopaedia Britannica*. 14th edn., 1966. Sections I, IV, V, VI and VIII. Reprinted, with revisions, up to 1973.
42 What should British Waterways policy be? 1—Pleasure cruising. *Modern Transport*. 97. No. 2462. March 1967. 29–31.
43 What should British Waterways policy be? 2—Commercial transport. *Modern Transport*. 97. No. 2464. May 1967. 44, 46–7.
44 Cruising on the Llangollen Canal. *Country Life*, 2 May 1968. 1136, 1139.
45 Telford, Jessop and Pontcysyllte. *Jnl. Rly. Canal Hist. Soc*. 15 (4). October 1969. 69–74.
46 Commercial transport: some possibilities. Chapter V in *Canal enthusiasts' handbook 1970–71*. Newton Abbot: David & Charles, 1970. (*See* Item 16).
47 Canals. Article in Georgano, G. N. (ed.). *A history of transport*. London: Dent, 1972. 197–214.
48 Trade and navigation. *IWA Bulletin*. 100. March 1972. 38–47.
49 Canal. Article in *Children's Britannica*. 3rd edn., 1973.
50 Transport. Article in *Children's Britannica*. 3rd edn., 1973. Section on Inland Waterways only.
51 The basic points [of IWA policy on commercial waterways]. *IWA Bulletin*. 104. March 1973. 44–7.

52 IWA's first Secretary – a tribute to Mr. L. T. C. Rolt. *IWA Bulletin*. 110. September 1974. 6–7.
53 IWA Continental visit – it was a huge success. *IWA Bulletin*. 115. December 1975. 27–9.
54 Lionel Munk – a distinguished IWA campaigner. *IWA Waterways*. 117. August 1976. 8–9.
55 Charles Hadfield records Fred's role [Fred Doerflinger as ISG Chairman]. *IWA Waterways*. 121. December 1977. 21.
56 The evolution of the canal inclined plane. *Jnl. Rly. Canal Hist. Soc.* 25 (3). September 1979. 94–101.
57 Glimpses of the North American waterways. *IWA Waterways*. 127. Winter 1979. 16–17.
58 Rivers and canals. Chapter V in Skempton, A. W. (ed.). *John Smeaton FRS*. London: Thomas Telford, 1981.

1.3   Forewords and Introductions

59 Introduction to
    THACKER, Fred S. *The Thames Highway Vol. 1: general history*. Newton Abbot: David & Charles, 1968. (Facsimile of original edition of 1914.)
60 Foreword to
    CLEW, Kenneth R. *The Kennet & Avon Canal*. Newton Abbot: David & Charles, 1968. 2nd edn. 1973.
61 Foreword to
    TEW, David. *The Oakham Canal*. Wymondham, Leics: Brewhouse Press, 1968.
62 Introduction to
    HARRIS, Robert. *Canals and their architecture*. London: Hugh Evelyn, 1969. 2nd edn. 1980 by Godfrey Cave, London.
63 Foreword to
    HOUSEHOLD, Humphrey. *The Thames & Severn Canal*. Newton Abbot: David & Charles, 1969.
64 Introduction to
    DE SALIS, H. R. *Bradshaw's Canals and navigable rivers*. Newton Abbot: David & Charles, 1969. (Facsimile of original edition of 1904.)
65 Introduction to
    PRIESTLEY, Joseph. *Historical account of the navigable rivers, canals, and railways of Great Britain*. Newton Abbot: David & Charles, 1969. (Facsimile of the 2nd edition of 1831, i.e. the 8vo edn.)
66 Introduction to
    PHILLIPS, John. *General history of inland navigation*. Newton Abbot: David & Charles, 1970. (Facsimile of the 5th edition, 1809).
67 Foreword to
    STEVENSON, Peter. *The Nutbrook Canal: Derbyshire*. Newton Abbot: David & Charles, 1970.
68 Foreword to
    HARRIS, Helen. *The Grand Western Canal*. Newton Abbot: David & Charles, 1973.
69 Introduction to
    GAYFORD, Eily. *The amateur boatwomen: canal boating 1941–1945*. Newton Abbot: David & Charles, 1973.
70 Foreword to
    ROLT, L. T. C. *The inland waterways of England*. London: George Allen & Unwin, 1979 (2nd edn.)
71 Foreword to
    HOUSEHOLD, Humphrey. *The Thames & Severn Canal*. Gloucester: Alan Sutton, 1983 (2nd edn.)

1.4   Reviews

72 COHEN, I. The non-tidal Wye and its Navigation. In *Jnl. Rly. Canal Hist. Soc.* 3 (4). July 1957. 69.

73 BRITISH TRANSPORT WATERWAYS (pub.). Inland Cruising Booklets: No. 3 Cruising on the Lee and Stort Navigations; No. 4. Cruising on the Staffordshire and Worcestershire Canal; No. 5 Cruising on the Shropshire Union Canal. In *Jnl. Rly. Canal Hist. Soc.* 4 (2). March 1958. 33.

74 CRANFIELD, J., & BONFIEL, M. Waterways atlas of the British Isles. In *Jnl. Rly. Canal Hist. Soc.* 12 (3). July 1966. 49.

75 WELSH, Edwin. The bankrupt canal. In *Transport Hist.* 1 (2). July 1968. 198.

76 CLAMP, Arthur L. Motoring and seeing waterways in Devon. In *Jnl. Rly. Canal Hist. Soc.* 15 (2). April 1969, 38-9.

77 D'ARCY, Gerard. Portrait of the Grand Canal. In *Transport Hist.* 3. 1970. 210.

78 SMITH, Peter. Waterways heritage. In *Jnl. Rly. Canal Hist Soc.* 18 (1). January 1972. 23-4.

79 ROLT, L. T. C. Landscape with canals. In *Newcomen Bulletin*. 109. December 1977. 9.

80 AMERICAN CANAL SOCIETY (pub.). The best from American canals and HAHN, Thomas F. The C & O Canal boatmen 1892-1924. In *Jnl. Rly. Canal Hist. Soc. Bk. Rev. Supp.* 37, November 1980. 7.

81 PASSFIELD, Robert. Building the Rideau Canal: a pictorial history. In *Jnl. Transport Hist.* 3rd series: 4 (1). March 1983. 91-2.

82 PASSFIELD, Robert. Building the Rideau Canal: a pictorial history. In *Jnl. Rly. Canal Hist. Soc.* 27 (7). March 1983. 209.

83 The Welland Canals; a comprehensive guide. In *Jnl. Rly. Canal Hist. Soc.* 27 (7). March 1983. 209.

2. Other Subjects

2.1 Books and Booklets

84 —— & ELLIS, C. Hamilton. *The young collector's handbook*. London: Oxford University Press, 1940. 2nd edn., entitled *The collector's handbook*, ed. by Charles J. Kaberry, 1951.

85 *Civilian fire fighter*. London: English Universities Press, 1941. 2nd edn. 1941.

86 D'AGAPEYEFF, Alexander, & ——
Maps. London: Oxford University Press, 1942. (Meridian Book Series); 2nd impression 1945; 2nd edn. 1950 (Compass Book Series No. 3).

87 WILMOTH, V. J. (Ed.)
Manual of firemanship (7 vols. in 8). London: HMSO, 1943-8. Hadfield wrote a significant proportion of this, but is not credited with the authorship.

88 EYRE, Frank, & ——
The fire service to-day. London: Oxford University Press, n.d. [1944]. 2nd edn. 1953.

89 —— & MacCOLL, James E. *Pilot guide to political London*. London: Pilot Press, 1945.

90 —— & MacCOLL, James E. *British local government*. London: Hutchinson's University Library, n.d. [1948]. (Hutchinson's University Library No. 14).

91 ALEXANDER, Charles. [Nom de plume].
The Church's Year. London: Oxford University Press, 1950. 2nd edn. 1956 (described as 2nd impression); 2nd impression 1959; 3rd impression *c*. 1963.

92 *Quaker Publicity*. London: Friends Home Service Committee. 1959. 2nd edn. 1962.

93 —— & HADFIELD, Alice Mary.
The Cotswolds. London: Batsford, 1966. 2nd edn. 1967 (described as 2nd impression).

94 *Atmospheric railways: a Victorian venture in silent speed*. Newton Abbot: David & Charles, 1967.

95 —— & HADFIELD, Alice Mary (eds. and contributors).
The Cotswolds: a new study. Newton Abbot: David & Charles, 1973. Paperback edn. 1981.

96 —— & HADFIELD, Alice Mary.
Introducing the Cotswolds. Newton Abbot: David & Charles, 1976.

2.2 Papers, articles and other short contributions

97 Diamond mining in the Transvaal. *Meccano Magazine*. 10 (12). December 1925. 689.

 98 MacCOLL, J. E., & ——
      The elector and the councillor. *The Fortnightly Review*. December 1945. 395–400.
 99 Common Service organisation in the central government. *Public Administration*. **28**. 1950.
      305–11.
100 The London Boroughs: Local Government in the shadow. *The Times*. 1 November 1950. 5.
      (Published anonymously).
101 Small town opens its diary. *Municipal Jnl.* **59**. 30 March 1951. 726, 735.
102 Fire-fighting to-day: tenth anniversary of former National Fire Service. *The Times*. 18 August
      1951. 5. (Published anonymously).
103 The Brussels universal and international exhibition 1958. *Jnl. Roy. Soc. Arts*. **106**, August
      1958. 681–94.
104 Fire-fighting. Article in *Children's Britannica*. 1st edn., 1960.
105 Quakers in business. *Wayfarer: the Quaker Monthly*. **40** (8). August 1961. 123–4.
106 The spirit in the family. *Wayfarer: the Quaker Monthly*. **42** (1). January 1963. 7–8.
107 Light in marriage [review of CALDERONE, Mary S. 'Release from sexual tensions'] . *Wayfarer:
      the Quaker Monthly*. **42** (8). August 1963. 116.
108 Review: FULLER, Edmund. The corridor. *Wayfarer: the Quaker Monthly*. **42** (12). December
      1963. 180. (Published anonymously).
109 A village experiment: three years of the South Cerney Trust. *Gloucestershire Life & Country-
      side*. February/March 1967. 14–15.
110 John Hogan. *The Quaker Monthly*. **48** (5). May 1969. 65–6.
111 A craftsman in words [review of HOGAN, J. P. 'One man's joy: essays'] . *Quaker Monthly*.
      54 (9). September 1975. 172.
112 The dangers of industrial archaeology. *WLIAS (West London Industrial Archaeological Society)
      Journal*. **10** (1). 1980. Np. Article 10/02/01.
113 CW at Amen House. *Newsletter of the Charles Williams Society*. No. 19, Autumn 1980.
      2–10.
114 April Fool's Day 1960 – and after. Article in *Good books come from Devon: The David &
      Charles twenty-first birthday book*. Newton Abbot: David & Charles, 1981. 5.

2.3   Edited

115 *A book of sea verse, chosen by E. C. R. Hadfield*. London: Oxford University Press, 1940.
      (Chameleon Book Series No. 11). 2nd impression 1946.
116 *A book of animal verse, chosen by E. C. R. Hadfield*. London: Oxford University Press, 1943.
      (Chameleon Book Series No. 21). 2nd impression 1949.
117 *Quaker Monthly* from 1963 to 1969.

3. Articles about Hadfield

118 [GORMAN, George H.] Charles Hadfield [as new editor]. *Wayfarer: the Quaker Monthly*.
      42 (1). January 1963. 6.
119 SHERWIN, Diana. Pride in a village: can other villages learn from the work of the South
      Cerney Trust? |*Farmer's Weekly*. 17 March 1967.
120 GORMAN, George H. Change of editor. *Quaker Monthly*. **48** (7). July 1969. 101.
121 DOERFLINGER, Frederic. Charles Hadfield – Hon. Consultant [to IWA]. *IWA Waterways*.
      118. December 1976. 6–7.
122 STREAT, Michael. Meeting people: Michael Streat| talks to Charles Hadfield. *Waterways
      World*. March 1978. 32–3.

CHAPTER 1

CANALS IN THE LANDSCAPE

THE VISUAL IMPACT of canals in the British landscape does not seem, at first inspection, to be very great. It would not be unreasonable to suggest that most people remain virtually unaware of their presence. The motorist turning down a side road might curse the steep hump of a canal bridge as he scrapes the bottom of his car on the tarmac; the traveller by train might be conscious that his route is following a remarkably similar track to that of the adjoining waterway; the traveller by air would be lucky to spot more than a brief, bright reflection of sunlight on still water. Yet, during the early years of the formation of the canal system, there was a very positive feeling that the canal engineers were in the business of changing the face of the landscape. How can one reconcile these two apparently contradictory views of the impact of the canals on the British scenery? Is it no more than a different perspective brought in by the passing of time? I hope to show that there is rather more to it than that: that there is a direct effect, a visual change brought about by the canal and its associated structures, and also an indirect effect, where change results from the introduction of a better, cheaper transport system.

The direct influence was the one which made the greater impact on contemporary commentators. Very early on in the canal age, the engineers introduced structures which seemed not merely dramatic but quite bizarre. The Bridgewater Canal had perhaps less importance as a piece of engineering than it had as a brilliant exercise in public relations. The praises that showered down on this waterway seemed to suggest that it was less a purely commercial concern, to be set against the latest turnpike road, than an artistic masterpiece, a *grand guignol* exercise in the picturesque comparable to the efforts of Capability Brown and William Kent or the newly-discovered romantic landscape of mountains and fells:

> 'Tis not long since I viewed the artificial curiosities of London, and now have seen the natural wonders of the Peak; but none of them have given me so much pleasure as I now receive in surveying the Duke of Bridgewater's navigation in this country.[1]

Certainly there were marvels introduced to the country by the canal builders. Barton aqueduct was but the first of many mighty structures which carried the new canals across the old river valleys: a process which culminated in the splendour of Pont Cysyllte. Yet even the mightiest are ignored by the great majority. Their day of fame has passed. Does this, then, mean that the effect of canals on our own 20th-century environment is strictly limited? Is the effect no more than a somewhat esoteric feeling experienced by canal travellers, a small-scale enjoyment of small-scale structures, set within the confines of a watery web spread across the face of Britain? Is the impact of canals on the landscape limited to a linear progression: does the influence never extend beyond the towpath boundaries? The short answer is – no, that is not the case. Yet it is true that the effect of canals on the land is less immediately obvious than the effects of more modern transport routes. The visual impact of a motorway is at once more dramatic and, I would argue, more damaging than that of canals. Indeed, the canal system built up during the second half of the 18th century and the beginning of the 19th century, could claim to be a rare, if not unique, example of a new, major transport system that enhanced rather than destroyed the surrounding landscape. It is nevertheless true to say that this aspect of canal development is now of comparatively minor importance, the appreciation of canal structures being limited to the minority who choose to explore the canals for pleasure. But this is far from being the whole story. It certainly gives little hint of the influence that the coming of the canals had on the development of both urban and rural Britain.

There is a very good reason why the canals themselves made so small a visual impact on the landscape and, paradoxically, when one investigates that reason one also gains a clue as to the way in which they were to act as conveyors of change. The structures that one finds along the canals – bridges, wharfs, warehouses and so forth – were often constructed as part of the whole process of building the canals themselves. The materials used in construction were, frequently, those of the country through which the canal passed. No better example of what this means in practice could be found than the southern section of the Oxford Canal, between Oxford and Napton. The canal begins its northern progress with a passage through the Oxford clay vale; it then passes on into a band of ironstone and oolite, which ends at Banbury with the arrival of the lias plain. These geological differences are reflected in the materials of the canal structures. The southernmost clays ensure the predominance of brick building. Brick, in this context, means brick fired in local kilns, resulting in a rich mixture of colours and textures in the brickwork, quite different from the drab standardisation that was to follow the establishment of the huge brickworks of a century later. As the canal penetrates the stone belt, so the structures reflect the underlying structures of the land; stone replaces brick as the primary building material, until brick again establishes a dominance on the lias plain.

All the early canal engineers relied heavily on local building materials. It was, after all, the high transport costs of such materials that helped bring the canals to birth. So, throughout the records of the canal companies, one finds references to the ways in which the engineers made maximum use of local materials, even when this involved setting up their own production units:

> We have burned some bricks at the foot of Devizes Hill . . . we have erected 3 Kilns in the best field & shall now have a regular succession for this time & good hard bricks.[2]

So, we find bricks being burned in an area well furnished with suitable clay, while in other regions special quarries would be opened up to take advantage of good, local building stone. As a result canal structures have become remarkably unobtrusive, originating in the landscape through which the canal passes, they have weathered over the years, so that they would seem to have blended back into their origins. In an age increasingly dominated by somewhat anonymous, mass-produced materials, we have come to appreciate this sense of belonging to a particular time and a particular place as a vital element in the visual appeal of the canal world. Yet this is a very modern attitude. A century ago, it was considered to be rather more desirable that a man-made structure should stand out from its surroundings. The highest praise was reserved for the grand, the gothic and, on occasions, the grotesque. The plain, simple and very local canal bridge was far from universally admired:

> Brindley did not sacrifice much money to architectural embellishments in these ugly brick-built bridges within the parish with which his silent highway is spanned.[3]

This criticism was made in 1865 by the historian of the Heyfords, villages strung along a stretch of the Oxford Canal which is now generally thought of as splendidly attractive and picturesque. Tastes have changed and will, no doubt, change again. But, however a new generation regards the Oxford Canal bridges, it is certainly true that the great majority of the population will never be aware of their existence. To a few they will remain as poignant reminders of a past when regional identity was to be seen displayed in every village and town, when buildings were an accurate reflection of the intrinsic nature of the ground on which they stood. But if the canals represented no more than that — an exercise in visual nostalgia — then their importance to the development of the modern world would be limited indeed. In fact, a second generation of canal engineers followed the pioneers and brought new notions and new methods of construction to the canal scene. If one returns to the Oxford Canal, one can obtain some idea of the ways in which the new canals differed from the old.

The northern section of the Oxford, from Braunston to Hawkesbury Junction, as originally laid out, followed the same sort of meandering line as the southern section and was similarly graced by structures built out of local materials. But there was considerable pressure on the company in the first part of the 19th century to improve this section, as it formed a vital link in the route from London to Birmingham via the Grand Junction Canal. So planning began in 1828 for shortening and straightening the line with new cuts across some of the more extravagant curves. The new straight line was achieved by building embankments over low-lying ground and by digging deep cuttings through rising ground. Some of the old curves were left as feeders, or as branches serving small wharfs. The towpath of the new canal had to be carried across the old on a series of bridges.

These old branches have all but disappeared, but one can still trace their progress through the fields at, for example, Newbold-on-Avon, where a typical early canal bridge has been stranded, canal-less in the middle of a field. One says typical, but in fact the design of such bridges was left very much to the whim of the individual contractor. Instructions to a local builder, John Watts, merely specified that he should construct 'two Wagon Bridges of good Brick coped with stone at the price of Two Hundred and Ten Pounds for both'.[4] The result was that these bridges on the old Oxford Canal were never standardised, but each differs in some degree from its neighbours. This is not true of the new line. The new bridges, that carry the towpath across the old line, no longer use old materials but the new building material of the Industrial Revolution — cast iron. The use of iron for bridge building had been pioneered by the Darbys of Coalbrookdale in the famous bridge across the Severn, completed in 1799. The canal engineers took this development a stage further — they introduced a form of mass production. A basic iron bridge was designed and cast in sections at the Horseley Iron Works, so that the same bridge could be repeated as often as necessary.

This idea of standardisation, of cost saving by mass production, was further developed on the canals and extended to other structures such as aqueducts. It was to prove very important to the later generation of railway engineers, who were also to adopt other techniques pioneered on the canals. The use of 'cut and fill' — digging cuttings and using the spoil for embankments — tunnels, the construction of skew bridges, and so on were all to play an important part in railway construction. It is interesting to note that the two engineers responsible for the Oxford Canal modernisation were Marc Brunel, whose son Isambard was to prove one of the great railway engineers, and Charles Vignoles, who undertook the survey and was later to work on many railway schemes including, if only for a time, the Liverpool & Manchester.

The chief engineer for the Liverpool & Manchester Railway was George Stephenson and, if one studies the line he built, it soon becomes very clear how he made use of canal engineering practice — and canal engineering thinking. The problems faced by canal and railway engineers were, in some respects, very similar, in that the canal engineers needed to provide a level route, and the railway men needed to keep to gentle gradients, in uneven country. Stephenson resorted to deep cutting and tunnelling at the Liverpool end of the line, and near Newton-le-Willows he dug a two-mile cutting, using the spoil for the adjoining embankment. Where the canal engineers built aqueducts he built the railway equivalent, viaducts, to cross the Sankey Navigation and the Bridgewater Canal. A skew bridge crosses the line at the most famous name on the railway, Rainhill, where Stephenson successfully ran his locomotive *Rocket* in the trials. And here, too, one can see how, in this early railway, canal thinking still obtained. Here is the railway summit level which, in canal terms, would have been approached at either end by locks. Stephenson followed a similar notion, building comparatively steep inclines at either end of the Rainhill level — the Whiston plane and the Sutton plane. It was believed that, at 1 in 96, these would be too steep for locomotives, and it was proposed to haul trains up by means of cables and fixed engines at the top of each plane. In the event, the locomotives coped without assistance, but this was not anticipated by the builders. A line laid down by a canal engineer would have followed a remarkably similar route to that of the railway.

The later waterways of the canal age had their spectacular features, the great aqueducts being the most obvious examples. Yet if one looks at what is perhaps the most spectacular British canal in terms of engineering panache, the Birmingham & Liverpool Junction, with its deep cuttings and high embankments, it is probably true to say that the visual effect is only experienced by those who travel it by boat. Its effect on the landscape is small and this could be said of the canal system as a whole. It is, quite literally, unobtrusive — far less obvious than the railways. So should one regard the canals as no more than the precursors of their 19th-century rivals — proving grounds for a new generation of engineers? It is certainly true to say that they fulfilled that role most effectively, and if they had done no more than set a pattern for others to follow, then their effect on the environment would have been considerable. But they are, in fact, rather more important than that, and contemporaries were very conscious of the ways in which the canals might be expected to bring changes to the community as a whole. A traveller to the Trent & Mersey Canal recorded one aspect of change:

> Notwithstanding the clamours which have been raised against this undertaking, in the places where it was intended to pass, when it was first projected, we have the pleasure to see content

reign universally on its banks, and plenty attend its progress. The cottage, instead of being half-covered with miserable thatch, is now secured with a substantial covering of tiles or slates, brought from the distant hills of Wales or Cumberland.[5]

This passage suggests that there had been a change in the style of individual buildings, following the spread of new materials. But when we try and investigate this effect in more detail, we find it very difficult to establish the facts with any accuracy.

We have already seen how the use of local materials helped to establish the identity of canalside buildings. In many areas, we find a marked contrast between the identity of canalside structures, such as bridges, lock cottages and warehouses, and other buildings surrounding the waterway. An excellent example can be seen at Fairfield Junction on the Ashton Canal. The canal passes through an area where the underlying rock is sandstone, and this stone is used for some fine canal bridges. But overlooking the junction is a massive 19th-century cotton mill — built not of stone but of brick. Indeed, brick soon appears as the primary construction material for the 19th-century buildings that line the canal. A similar story can be seen repeated on a larger scale in another cotton area, Blackburn. The locks on the Leeds & Liverpool Canal are surrounded by stone stables and warehouses, while stretching away from them are the 19th-century mills that line the bank, and they are again built of brick. This would appear to offer confirmation of the theory that the canal age was a period when cheap building materials were moved around the country, beginning a process whereby local materials and styles were to be submerged under a new uniformity. The story is not, however, quite that simple. Closer inspection of the buildings reveals that most of the mills are more recent than the canal structures and fall within the railway age. They do not, of themselves, provide convincing evidence that change, in the shape of brick to replace stone, came by canal. In any case, bricks could be brought from comparatively close at hand.

The two most significant changes in building patterns, as far as the appearance of individual buildings was concerned, were the gradual disappearance of thatch to be replaced by slate, and the increasing dominance of the brick-built house, using standard mass-produced bricks. The question to be answered is — to what extent did these changes depend on the new, low-cost transport offered by the canals? Slate certainly featured heavily in the list of cargoes carried on the Lancaster Canal. The Lake District had a thriving slate industry, and trade on the canal was improved by the construction of the arm from the main line to the new port of Glasson. The first sea-going vessel to make use of the new branch was the sloop *Sprightly* in May 1826 with a cargo of slate for Preston.[6] The proprietors of the short Ulverston Canal considered the slate trade sufficiently important for them to go to the extra expense of constructing a road to ease the traffic from quarry to canal.[7]

The other great centre of the slate trade which was to develop enormously during the 19th century was North Wales. Few of the slate mines and quarries in this mountainous region were conveniently placed for a canal. The Montgomeryshire Canal with its connection to the Ellesmere and through that to the vast Midlands network was, however, well placed to take advantage of the trade. As Priestley wrote in his canal survey of 1831:

The northern parts of the line in particular, and the whole line generally, is situated in the midst of quarries of limestone, slate and freestone ...[8]

The next entry in Priestley, however, gives an idea of where the slate trade would increasingly look for transport. It refers to the Nantlle Railway, described in its enabling Act of 1825 as 'a Railway or Tramroad, from or near a certain Slate Quarry called Gloddfarlon, in the parish of Llandwrog, in the county of Carnarvon, to the town and port of Carnarvon'. Railways and tramways were to find increasing use in carrying slate from mines and quarries. They did not, however, take slate to the customers, but to ports such as Porthmadog for shipment round the coast and much of this slate would, in the pre-railway age, have continued its journey by inland waterway. So it was that the 'miserable thatch' of a Staffordshire town came to be replaced by Welsh or Cumbrian slate.

The increased use of slate was widely welcomed, though not by all. The romantics even then deplored the removal of the old 'picturesque' thatch, a view echoed today by those who enjoy the sight of a well-tended, thatched roof. But, today, we are usually seeing first-class thatching, using the best materials, not a soggy, dripping mess of wretched straw and heather. Slate brought new standards of comfort to many homes, though it also helped in the spread of uniformity throughout Britain.

The spread of uniformity was carried even further by the introduction of stock bricks. The canal companies certainly regarded bricks and brick tiles as likely cargo, for they were listed in the tonnage rates of many concerns. Sometimes they appear as simply one item in a long list, covering a wide variety of commodities, as with the Regent's Canal — 'Lime, Lime-stone, Chalk, Bricks, Tiles, Slates, Lead, Iron, Brass, Copper, Tin, Platina, Stone and Timber of every Kind' at a rate of 8d. per ton as far as Mile End.[9] On a few waterways, however, such as those of the Humberside region, where bricks would be of much greater importance, costs were more scrupulously listed. The Ancholme Navigation not only separated them out from other building materials, but quoted tonnage rates at 1s. 8d. per thousand, rather than by weight. But how far was this a purely local trade and how far were the bricks carried around the country? Brick building was, after all, common in many areas before the canal age, but the bricks were locally fired from local clays. This was of course, not possible in all areas, as the proprietors of the Oxford Canal Company discovered during their work on the improvement of the northern section. They dug clay pits and fired their own bricks, the so-called 'white bricks'. These proved to have a disastrously short life and were rapidly either replaced or faced with a more durable brick, mostly the red brick of Oxfordshire. Nevertheless, local bricks were common in the 18th century, and they came in a wide variety of colours and textures, reflecting the differences in raw materials and the somewhat hit-and-miss firing techniques in use at the kilns. The result was a great richness in the buildings using such bricks. Standardisation wrought changes, so that the term 'red brick' came to be used in a rather scornful sense. This is the great change with which we are concerned, not the use of bricks as such, which was already widespread in the 18th century, but the use of standardised bricks. How far was that process helped by the spreading canal system?

In this instance, the answer must be — not a great deal. Brick carrying was of minor importance, and this is not too surprising. The introduction of large-scale brick manufacture came comparatively late onto the scene, encouraged by the introduction of the Hoffman kiln in the 1850s. The canals were not, by then, very well placed to compete for a new traffic. Nevertheless, fleets of brick boats were used to carry the stock bricks from the brickfields into London. It was considered a very poor grade of cargo. The boatmen engaged in the trade were badly paid and the boats were of the poorest quality. The general nature of the trade was not improved by their usual return cargo of manure or refuse from London's ash pits.[10] The influence of canals in changing British building practice could best be described as small but not negligible.

We have already seen one example of a contemporary tourist's view of the canals as bringers of change. Here is a second quotation, again relating to the Trent & Mersey, and indicating a different method of hurrying through the processes of change:

> And though the expence attending this astonishing work was enormous, so as to promise little or no profit to the adventurers; yet in a few years after it was finished, I saw the smile of hope brighten every countenance, the value of manufacture arise in the most unthought of places; new buildings and new streets spring up in many parts of Staffordshire, where it passes; the poor no longer starving on the bread of poverty; and the rich greatly richer. The market town of Stone in particular soon felt this comfortable change; which from a poor insignificant place is now grown neat and handsome in its buildings and from its wharfs and busy traffic, wears the lively aspect of a little seaport.[11]

Here we have the presence of a canal not so much affecting the nature of individual buildings as changing the overall pattern of development. Returning to the Oxford Canal one can see what this meant to a predominantly rural area. At various points along the route there is an interaction between the existing geographical features, the new canal and other transport routes. A typical example is to be seen at Enslow where what is now the B4027 crosses the canal. Here one can see the evidence of a little flurry of building activity with wharf, houses and an inn. Today, it might seem a somewhat isolated settlement, but a glance at the pub sign gives the first hint as to how and why the settlement was generated. The pub is the *Rock of Gibraltar*, and it is the first word that provides the hint. A little investigation soon reveals that this is an area liberally supplied with small stone quarries, and stone was just the sort of bulk cargo on which the canals thrived. A closer look at the road reveals that here, too, all is not quite what it seems. The grass verges are exceptionally wide and, although it has been joined by a modern concrete structure, the original stone bridge still stands and is wide and handsome. In fact; this secondary road was once the main London to Worcester coach road. Put all the evidence together, and it is not surprising that the conjunction of main road, canal and a source of cargo should give rise to a small settlement. Although such a settlement remains a minor feature in the landscape, it provides an indication of the way in which junctions, of themselves, generate building activity. The more important the junction, the greater that activity will be. The most important junctions on the canal system were those where broad waterways and narrow canals met. These became trans-shipment points where goods were stored awaiting exchange between narrow boats and a variety of broad-beamed craft. Such interchange points often developed into large complexes and gave rise to the major canal towns, such as Shardlow, Stourport and Ellesmere Port, which owed their entire existence to the coming of the canals. The importance of these towns has been stressed by many writers, such as Professor Hoskins who described Stourport as 'the emporium for the west Midlands',[12] and it would sometimes appear that the influence of canals on urban development begins and ends with these classic towns. This is very far from being the case. It could, indeed, be argued that wherever a canal met an existing town, it had some influence on the way in which the town grew. This soon becomes evident when one begins to look at a few towns in more detail.

The obvious place to start is the industrial town. Such towns grew up during the period we call the industrial revolution, and this is precisely the time when the canals, too, spread across the face of the land. If there is to be any evidence of a relationship between town development and canal building then this is where one would expect to see that evidence at its clearest. A suitable starting point for investigation would be the Huddersfield Narrow Canal, joining as it does Lancashire cotton at one end with Yorkshire woollens at the other. Walk out along the canal from Huddersfield, and it is at once obvious that the waterway has formed a major line of development for the woollen industry of the 19th century. On the hills above, and in small hilltop villages such as Golcar, one can see ample evidence of the older form of cloth manufacture in the many cottages of the hand-loom weavers. Here in the valley, however, the great mills dominate the scenery. It would be simple to draw an obvious conclusion: the canal was built, and the industrialists followed to take advantage of the improved transport facilities. Certainly there is no lack of physical remains to support that view. At the Huddersfield end, the gaunt, tall mills stand with their walls sheer to the water, creating an artificial canyon, broken only by the spaces left for coal wharfs. There is, however, a complicating factor to this simple story of cause and effect. The canal shares the valley floor with the River Colne, and that too provides a natural line of development. There were mills here using the river as a source of power before the canal was built. The map of the proposed canal, produced by the company, shows an area already well developed as an industrial region, with mills along the river all the way from Marsden to Huddersfield. Even quite minor tributaries had their share of mills. Thorn's Clough, a small stream running from Standedge Foot to Marsden, had four mills inside a mile, and there are a further three mills on Car Brook.

Looking at such early maps, it would now seem that it was the presence of mills that attracted the canal promoters in the first instance. The story might be adjusted to read: mills were built along the Colne and tributaries; the presence of the mills suggested the need for a canal; the canal was built and encouraged the building of yet more mills. There is certainly evidence that this was indeed the case. Many of the later mills, powered by steam rather than water wheel, were placed alongside the canal to take advantage of cheap coal deliveries. In Slaithwaite one can see how both river and canal formed axes of development. Mills were built on both banks of the river, and mills line the canal on its passage through the town. The canal certainly had an effect on the siting of the mills, but it could be argued that the mills would have been built even if the canal had never come to the Colne valley, and natural geography would have ensured that they would have been sited much as they are today.

If the constrictions and contours of the Colne valley make it difficult to gauge the precise effect of the canal on the pattern of development, perhaps other areas might give more direct evidence. Travellers on the Leeds & Liverpool Canal as it passes through the Lancashire cotton area can hardly fail to be aware of industrial development along the way, just as with the Huddersfield. Here, however, patterns are rather different. Take, for example, the town of Blackburn. A map of 1825[13] shows a very precise pattern of building. Two rivers pass through the area — the Darwen and the Blackwater — and both have had their effect on development. But here the canal

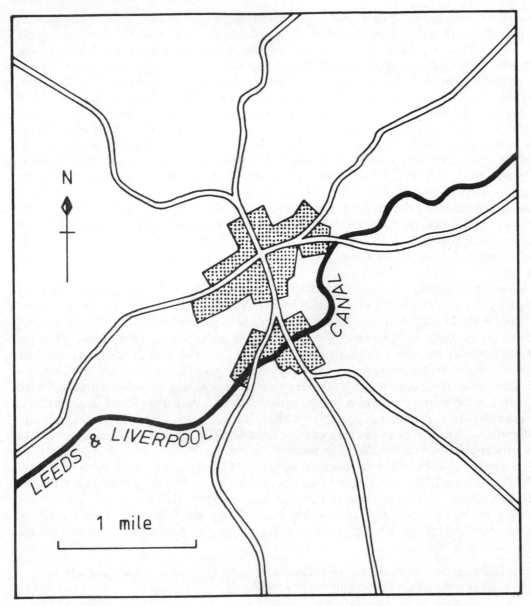

Fig. 1.1. Blackburn, 1824: the buildings along the canal consist of mills and
associated terraced housing.

has not followed the obvious river route; indeed, it crosses the Darwen on an aqueduct.
So there is a very definite separation between riverside development and canalside
development, and this shows clearly on the map. When built, the canal followed a line
well to the south of the town centre, and a new core of development can be seen
where the roads from the town cross the canal.

By 1847, when the Ordnance Survey map appeared, the effect of the canal could
be seen to be increasing. The Blackwater still formed a relatively important axis for

industrial expansion, with spinning mills such as Water Street mill, power-loom factories and, surrounding these, terraces of back-to-back houses. But activity along the canal is even more marked. There are, as one would expect, a number of wharfs and warehouses, especially grouped around the locks, but there are also a number of large mill complexes, such as Enam Mill, with both spinning and weaving. And factories now spread out all along the waterway – Navigation Mill, Canal Foundry, Canal Mill and so on. And where the mills could not squeeze into a space with a water frontage, as with Daisyfield Mill, set back some two hundred yards, a new road, Canal Street, was built to join mill and canal. Other industrialists opted for a canal site, but were forced to move way out of the town. The steam-powered spinning and weaving mill, Green Bank Mill, is a good example, and two terraces of cottages, Crown Street and Abbot Street, were built for the mill workers. Blackburn provides strong evidence for the importance of the canal in determining the direction of development for the town, and similar evidence can be obtained from other cotton towns. In Rochdale, for example, a map of 1824[14] shows the canal basin with its extensive wharfs and warehouses forming a new nucleus for development, well separated from the old centre of the town.

In a very few cases, canalside developments were on a much larger scale and were initiated by an individual. In the textile world, Saltaire, the mill town built by Sir Titus Salt, with its huge mill straddling the Leeds & Liverpool Canal, is a fine instance. But one can go back to the very beginning of the canal story for a classic example of an industrialist founding both factory and town on the line of a new canal. The famous potter, Josiah Wedgwood, was the principal promoter of the Trent & Mersey Canal, and from the very first his advocacy of the canal was closely tied to his own dream of a new pottery complex bigger than any that had ever been built before. As a manufacturer producing pottery in the classical style, he selected a suitably classical name for his new development – Etruria. The building of the canal and works proceeded simultaneously, the success of the latter being very dependent on the former. Indeed, others soon became aware that Wedgwood stood to gain a good deal from the canal, and even started rumours that he had used his influence on the Canal Committee to change and distort the line of the canal in the region for his personal gain. Wedgwood, who had toiled mightily to get this pioneering waterway begun in the first place, was hurt and saddened by the rumours. He wrote to his partner

> This cloud has been gathering for some time, & no pains are spared by the Party who have blown it up, to make it light as heavy upon me as possible.[15]

The criticisms were unjust, but they give some notion of the importance attached to canal transport by early industrialists. Etruria has all but vanished, yet it was the most important works to be built in the potteries in the 18th century, establishing a pattern of building that was to be followed for the next hundred years. And here, at least, one can say quite unequivocally that the canal played a vital role.

Of all the towns and cities where development can be said to have been vitally affected by the canals, there can be none where the effect was more pronounced than Birmingham. As with the pottery towns that now make up Stoke-on-Trent, Birmingham had the natural ingredients for industrial development, but was some way from a navigable river, and thus lacked a suitable transport system, so essential to industrial growth. The canals provided that system. Tracing the expansion of Birmingham in the second half of the 18th century through old maps[16] one's first reaction is astonishment at just how rapid that expansion was. Then one soon becomes aware of how the direction of expansion was determined by the spreading network of canals. If one compares Samuel Bradford's map of pre-canal Birmingham in 1750 with Thomas Hanson's map of 1778, one finds that virtually all the growth has been canal-orientated. The new Birmingham Canal comes in from the east, terminating in a wharf. The whole area between the new wharf and the old town is filled in with houses and commercial development, centred round the appropriately named Navigation Street. By the end of the century, heavy industry has become a marked feature of the Birmingham scene, and the spreading network of canals forms the link. The Birmingham & Fazeley is already, by this date, lined with new industries, including breweries and steam mills, while the original Birmingham Canal has become the focus for huge expansion, with brass houses and iron foundries and the city's most famous works, the Soho foundry of the pioneers of steam engine construction, Boulton and Watt. The canals seem like veins in the body of Birmingham, enriching the city and stimulating growth. They continued to stimulate in the 19th century, but could no longer contain it. Birmingham was now set on its way to becoming the second most important city in England. It spread in every direction, and those directions no longer depended on the still extending canal system. Yet it is true to say that without the canals that growth would have been held back for a generation or more, and there is no reason to suppose that it would have followed the same overall pattern.

It is in the industrial areas that one looks for, and indeed finds, the strongest possible evidence to support the theory that the canals played an important part in establishing patterns of urban development. It must, however, be remembered that other industrial centres developed with no help whatsoever from the canal system. One should, then, be cautious about overestimating the importance of the canals. It would seem to be true to say that the industrial revolution developed in good measure on the cheap transport offered by the canals, and in those areas where the canals were built they had a decisive effect on the pattern of change. In other cases, one might expect to find the canals to be rather less influential. Yet even if one looks at towns as far removed from the industrial world as it is possible to get, spiritually if not geographically, one finds that the arrival of a canal in an area was far from unimportant. Look, for example, at Royal Leamington Spa.[17]

A map of 1783 shows plain Leamington, no spa and certainly not royal. It was, in fact, a small rural community, thinly settled around the crossing of the river Leam. The nearest thing to a coherent development is a terrace of cottages in Bath Lane, described as 'occupied by Mr. Wise's labourers'. By 1818, however, Bath Lane had been aggrandised into Bath Street, lined with bath houses and further extended into Clement Street and Brunswick Street. The area to the north of the river had now also become important, with the Pump Room, new Assembly Rooms and a hotel. The transformation of Leamington into Leamington Spa was firmly established, and this is the handsome Georgian town one can still see today. The canal too had arrived, but was very much shunted away around the southern edge of the fashionable spa — kept, as it were, at arm's length. The map of 1835 shows a somewhat different picture. The

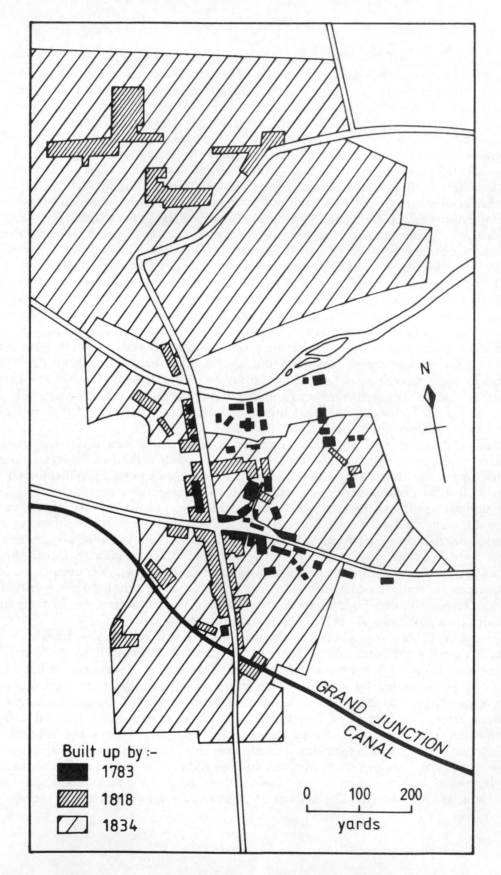

Built up by :-

■ 1783
▨ 1818
◻ 1834

0   100   200
yards

Fig. 1.2. Leamington: the 1834 development around the canal consists almost entirely of factories and narrow terraces. The grander houses connected with the fashionable spa are in the northern development.

GRAND JUNCTION CANAL

N

spa still goes its own way, but now a new Leamington is developing around the canal: a town of narrow terraces in marked contrast to the wide streets and squares of the spa. Twelve years later this new area shows other signs of activity. The town gas works has now been established on a canalside site, and industry has appeared in the shape of Smith & Taylor's Foundry. These two Leamingtons have continued to develop side by side right up to the present day, so that when one travels the Grand Union Canal through Leamington one still gets scarcely a glimpse of the elegant spa.

The canals can so easily get lost in the modern landscape. This is part of their charm; this is the quality which makes them ideal for leisure cruising. To say that, however, is not the same thing as to say that their influence on the landscape has been negligible. In their role as major transport routes of the industrial revolution, they were vital to that greatest of all periods of change. This is true in the general sense that they quickened the pace of change, and I have also tried to show that this is true in a more particular sense as well. Wherever canal and town have met, the canal has changed the town in ways which are still visible today. The canals may long since have ceased to play a major role in the transport of the country, but the land still carries the permanent mark of the days of their greatness.

NOTES AND REFERENCES

1. ANON. *The history of inland navigations.* 1766.
2. Engineers' reports to the Kennet & Avon Management Committee, 27 January 1803.
3. WING, William. *Annals of Heyford Warren.* 1865.
4. Oxford Canal Co. Committee Book, 2 January 1770.
5. PENNANT, Thomas. *The journey from Chester to London.* 1782.
6. HADFIELD, Charles & BIDDLE, Gordon. *The canals of North West England.* Vol. 1. 1970. 196.
7. *Ibid.* 208.
8. PRIESTLEY, Joseph. *Historical account of the navigable rivers, canals and railways, throughout Great Britain . . .* 1831.
9. *Ibid.*
10. HANSON, Harry. *The canal boatmen 1760–1914.* 1975.
11. SHAW, Rev. Stebbing. *A tour to the West of England in 1788.* 1789.
12. HOSKINS, W. G. *The making of the English landscape.* 1955.
13. JOHNSON, William. *Map of northern canals.* 1825.
14. *Swine's survey of Rochdale.* 1824.
15. Wedgwood to Thomas Bentley, 10 April 1768.
16. BIRMINGHAM PUBLIC LIBRARIES. *Birmingham before 1800.* 1968.
17. All Leamington maps referred to are held in the Leamington Public Library.

1. Charles Hadfield.

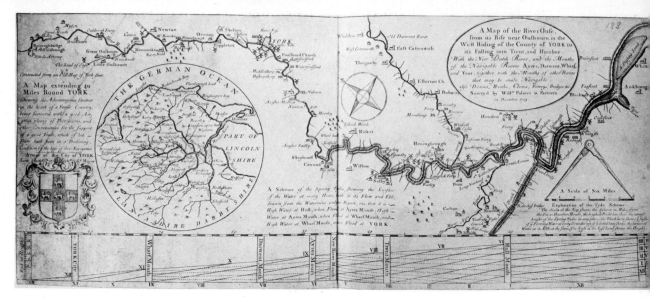

2. Part of William Palmer's map of the Yorkshire Ouse, 1725.

3. Stanch on the Little Ouse at Santon: painting attributed to F. W. Watts (early 19th century).

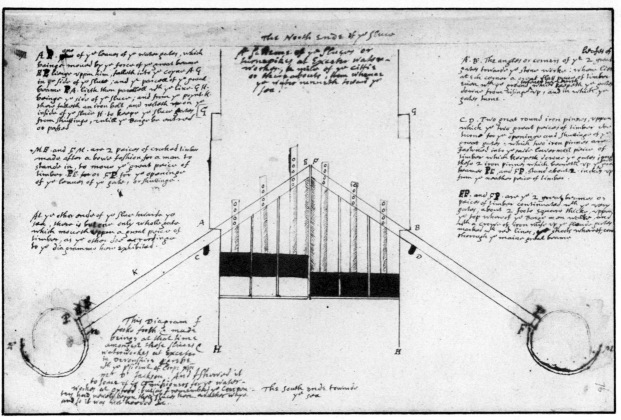

4. (*above*) Exeter Canal lock gates: drawing of c.1630.

5. (*below*) Denver Sluice, as completed 1750: drawing of c.1810.

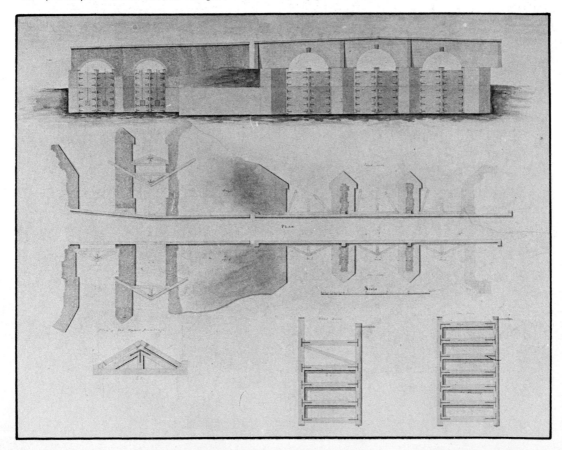

6. (*left*) A breakthrough for push-tow operators on the Rhine would be the long-awaited authorisation o six-lighter tows in the Netherlands (such tows are at present limited to the German Rhine between Emmerich and Koblenz). A test run with such a formation, using the push-tug *Mannesmann IV*, is here seen on the Hartelkanaal, which parallels the main shipping channel, the Nieuwe Waterweg, between Rotterdam and the North Sea.

7. (*below*) The *Maba Kelce* with a 49-barge tow on the Mississippi.

8. (*right*) A British Waterways Board push-tow passes two Cawoods Hargreaves push-tugs at Castleford on the Aire & Calder Navigation.

9. (*below*) One of the first push-tugs brought into use on the Rhine was the *Herkules*, built about 1959 to an unusual design developed by Dr. Westphal in association with the Ruhrort shipyard and engine factory (part of the Krupp concern). It had catamaran hulls 5m. distant, each with its own 620h.p. engine and rudders, giving reduced resistance and greater manoeuvrability thanks to the large distance between the screws. The *Herkules*, 26.80m. by 15.00m., is here seen on the Waal at Zaltbommel, with its 10 lighters distinctly resembling the Tom Puddings operated on the Aire & Calder Navigation. Unlike the latter, however, these lighters, 15m. square and loading 500 tonnes, could not be automatically unloaded by a hoist, and the principle was abandoned in favour of larger lighters.

10. A 4,000-tonne push-tow negotiating an awkward stretch of the Meuse in Liège, during an official trial run in 1981. The trial (of two Europa class lighters, each 76.5m. by 11.4m.) proved that the operation of such tows was practicable on the Meuse.

11. & 12. Architectural vandalism in Birmingham: the demolition of the church represents a loss for the canal environment.

13. Farmer's Bridge Locks, Birmingham: 'De Maré introduced me to a friendly unassuming world, an environment that was straightforward . . . but a scene that asked questions'.

14. Broad Street Tunnel, Birmingham: enclosure creates a secret world.

5. Hartshill, Coventry Canal: landscape and waterside wild flowers combine with simple buildings to create a magic appeal, often 'entirely conceived and designed without an architect in sight'.

16. 'We see neglect and indifference and attitudes that seem to dump the community's prejudices into the canal itself.'

17. This drawing inspired the creation of the James Brindley Walk Scheme, designed by Peter White in 1967. Old buildings were conserved and converted, and a new public house and waterside houses introduced around wharves. Birmingham discovered that canals could have an exciting role in the provision of amenity and leisure in inner city areas.

18. Planning neglect and land use indifference are serious problems for the canal network. The development of new attitudes and the attraction of investment to counter these problems are important objectives. By encouraging local authorities to take advantage of water-related sites—not to treat canals as awkward boundaries—much can be done to enable our waterways to provide positive benefits to the millions who live and work by them.

19, 20 & 21. Simple and timely maintenance both emphasises the quality of buildings, and reduces vandalism. 19: (*left*) lock and toll office at Bratch on the Staffordshire & Worcestershire Canal. 20: (*above*) footbridge on the Birmingham & Fazeley Canal. 21: (*below*) Stretton Aqueduct on the Shropshire Union Canal.

22, 23 & 24. One way of maintaining the tradition and value of old buildings is to convert them sensitively to new uses, as these examples show. 22: (*top left*) stables at Sneyd Yard, Birmingham Canal Navigations. 23: (*bottom left*) warehouse and workshop at Apperley Bridge, Leeds & Liverpool Canal. 24: (*above*) a Telford Cottage at Oldbury Yard, Birmingham Canal Navigations. All have been converted to British Waterways Board offices.

25. The attractive woodlands bordering many of our waterways need imaginative forestry management and design if their ecological interest is to be preserved.

26. The co-ordination of normal maintenance with special measures for enhancement, as at this award-winning site (Hawkesbury Junction), can attract local authority investment for picnic areas and other facilities for the public.

27. (*left*) The first floor of an interesting early 19th-century Oxford Canal workshop building at Hillmorton is now Peter White's office.

28. (*below*) Effective influence over waterside planning applications is essential, otherwise a mediocre waterway environment can be created by, for instance, untidy storage and parking areas.

29. & 30. Efficient modern structures need not be characterless; skill and imagination can create a more pleasing result which is also cheaper to maintain. Compare the crude brick felt-roofed box above (29) with the more successful structure below (30).

31. Hawkesbury Junction—a conservation area developed jointly by BWB and the local authorities.

32. Members of the Pocklington Amenity Society at work restoring Baldwin's Bridge, Pocklington Canal.

How about helping repaint a canalboat, rebuild a 300 year old watermill, or restore life and beauty to a dry canal?

Of Britain's thousands of miles of inland waterways, long stretches are in desperate need of some kind of restoration. And each weekend all over Britain groups of dedicated enthusiasts hack, weed, burn and dig, taking the task upon themselves.

In 1977 Shell and the Inland Waterways Association started a scheme to assist. We couldn't fund entire projects, but we were often able to help get one started, or completed.

Last year, for instance, we supporte[d] 92 different projects as diverse and unus[ual] as starting a water adventure playgroun[d on] the Water of Leith, and rescuing a forlor[n]

Fancy h
weed a

33. An appealing advertisement publicising one of Shell's previous award schemes.

...genarian canalboat from the clinging ... of Foxton barge-lift.

Other continuing projects include ...vating the famous Caen Hill flight of

29 locks on the Kennet & Avon Canal, and restoring the venerable steam engines of Ellesmere Port Boat Museum.

All these, and many other projects, richly deserve help and support. More, in fact, than we are able to give.

The problem is vast, but luckily it's a problem that's fun to help solve.

If you'd like to help, or would like more information about Shell's community affairs projects, write to Hazel Barbour, Community Affairs, Shell UK Ltd, Shell-Mex House, Strand, London WC2R 0DX.

And next weekend you could be weeding riverbeds instead of flowerbeds.

...lping ...iver?

You can be sure Shell's playing its part

34. (*above*) Lord Greenwood of Rossendale opens the rebuilt Flatford Lock on the River Stour, Easter 1975.

35. (*left*) Lock rebuilding on the Basingstoke Canal at St John's, Woking.

CHAPTER 2

ENGINEERING ON THE ENGLISH RIVER NAVIGATIONS TO 1760

THIRTY YEARS AGO, under the inspiration of Professor T. S. Willan, I wrote a paper on the engineers of the early river navigations in England. Since then much more has been learnt and I welcome the opportunity of presenting a new synopsis on the subject, but this time with emphasis on the engineering works. The engineers will of course be mentioned, with some corrected or recently discovered dates. Biographical details, however, must await a future publication.

What is needed, and attempted here, is a concise chronological account of the more interesting and important works from Elizabethan times to the mid 18th century; before, that is to say, the beginning of the 'Canal Age'. A complete account even up to this period would exceed the space available, while another chapter (at least) would be required for a proper treatment of river engineering after 1760.

Willan's classic monograph of 1936 is still the best general treatment of early river navigation in England. Histories of individual navigations are admirably set out in the relevant volumes in Charles Hadfield's series *The Canals of the British Isles*, and are summarised by Paget-Tomlinson* who also explains most of the engineering terminology involved. It can be taken for granted that where more specialised references are not given, historical facts have been taken from Hadfield for rivers in the South, the Midlands and North-East, from Hadfield & Biddle for the North-West, and from Boyes & Russell for Eastern England.

Extent of the navigations

The figures given in Table 1 exclude the lengths of tidal rivers navigable by sea-going ships: the Thames to London Bridge, the Avon to Bristol and so on. Important as they are, our concern is with inland waterways navigated by river craft.

On that definition, there were in England from time immemorial rather more than 650 miles of open, or natural, river navigation. York, Nottingham, Peterborough and Maidstone are examples of towns marking the heads of such navigations. In addition, before 1600 a further 90 miles (approximately) had been added by engineering works. This includes the tiny Exeter Canal with its Elizabethan pound locks, but chiefly comprises the Thames from Maidenhead to Oxford and the Lee from Bow to Hertford on which, in both cases, navigation was possible by means of flash locks. Most of the flash locks, but not all, were in mill weirs.

* *The complete book of canal and river navigations.* Waine Research, 1978.

Systematic efforts to improve, or rather to extend, the river navigations were made in the 17th century. During that period the length increased by nearly 200 miles, involving the construction of at least 50 pound locks and numerous weirs, sluices, flash locks and new cuts.

In the next 60 years 370 miles were added to the total length. Or, to put it another way, by 1760 about 700 miles of navigation had been created or improved by engineering works; and this at a time before any of the major English canals were built.

The maximum extension of the river navigations can be dated very roughly to 1815, a date chosen partly because it marks the completion of an important episode of lock building on the Thames. The total length then amounted to almost 1,600 miles. Few of the bigger towns, other than those in the heart of the Midlands, were now inaccessible to navigable river water, fluvial or estuarial.

TABLE 1

Mileage of River Navigations in England

	1600	1650	1700	1760	1815
Natural	670	670	650	620	580
Engineered	90	150	300	700	1,000
Total	760	820	950	1,320	1,580

TABLE 2

River Gradients

River	From – To	Length miles	Fall feet	Average Gradient ins./mile
Thames	Lechdale – Teddington	129	219	20
Great Ouse . . .	Bedford – St Ives	31	65	25
Aire.	Leeds – Weeland	35	78	27
Nene	Northampton – Peterborough	61	180	35
Lee.	Hertford – Bow	31	111	42
Wey	Guildford – Weybridge	20	72	43
Kennet.	Newbury – Reading	23	128	66
Calder	Sowerby Bridge – Wakefield	25	188	90

Elizabethan navigation works

Exeter Canal   No records have yet been found of any pound locks in England earlier than those on the Exeter Canal, built by John Trew of Glamorganshire who was employed as engineer-contractor by the city of Exeter.[1] Construction of the canal began in February 1564. It was a cut 1¾ miles in length from Exeter to a point on the tidal river Exe about two miles above Topsham at the head of the estuary.

A timber weir across the river maintained an adequate depth of water above the cut, and water level in the lower part was retained by a sluice. As this 'opened by the tyde' it presumably had a self-acting pivoted gate or perhaps a pair of mitre gates pointing upstream. The canal, embanked for some of its length, had a width of 16 feet and a depth of three feet suitable for boats or lighters up to 16 tons burden.

There were three turf-sided locks, the lower two being 189 feet in length and 23 feet wide, with a 5½ft. depth of water, and therefore large enough to contain several vessels (a familiar feature of early pound locks in Europe) and to act as passing places. The upper lock was considerably wider, to serve as a basin or dock. The navigation appears to have been opened in the autumn of 1566 and the whole job, including a masonry quay wall 150 feet long on the riverside at Exeter, was completed in 1567. Contemporary documents refer to the lock chambers as 'pools' and the lock gates as 'sluices', so there were seven 'sluices' in all. Some improvements took place in 1581 but thereafter the navigation remained essentially unchanged for nearly a hundred years.

A drawing (Figure 2.1) dating from *c*. 1630,[2] of 'the Sluices or turnpikes at Exeter water-workes' shows the upper gates to be mitred, with paddles and pivoted balance beams. The stems of the paddles pass through holes in the beams and are levered up a few inches at a time using a crowbar and iron pins. A note records that at the lower end of the lock 'there is but one only whole gate which moveth upon a great piece of timber, as the other doe accordinge to the diagramme here exhibited'. These details may apply to the 1581 improvements rather than the original structure, but can in either case be taken to represent an example of Elizabethan engineering.

Lee   Commissioners were appointed in 1575 to improve the Lee navigation. They gave orders for scouring the river, laying a towpath and regulating the use of the flash locks. There were probably at least a dozen of these between Tottenham and Ware, and some of them seem to have been of the guillotine type, known on the Lee as 'turnpikes'. However, the Commissioners also ordered the construction of a pound lock on a new cut 200 yards in length at Waltham mill, to take the place of a nearby turnpike which was to be transformed to a weir. Built in 1576, with timber walls and floor, the lock definitely had two pairs of mitre gates enclosing a chamber about 60 feet long by 22 feet wide and 10 feet deep. Specifications for the timber work have survived, and the total cost including the cut came to £272.[3]

Serious settlements necessitated reconstruction of the lock on masonry foundations only three years later. This operation, completed in June 1579, was directed and supervised by a 'Mr. Trew': surely John Trew of the Exeter Canal.

Local intrigues led to illegal demolition of the lock in 1592, and the turnpike was reinstated by the mill owner who charged tolls for its use. Despite a life of only sixteen years the historical importance of the Waltham pound lock has long been recognised as the first 'modern' example in England.

Shelford Lock   Details of a rather more primitive lock of about the same date, *c*. 1576, have recently come to light.[4] Built by a mill owner and opposed by the river men, it was situated in a new cut alongside the Trent at Shelford and is shown, together with a long weir, on a plan of 1592. The gates, referred to as 'The Lockes' on the plan, are placed about 75 feet apart enclosing a chamber no wider than the cut itself. They are of the guillotine type with posts carried upwards to an overhead beam or lintel, and there are two wheels for lifting or lowering each gate.

Thames   It was in 1585 that John Bishop compiled the first complete list of locks and weirs on the Thames.[5] There were in fact no locks from London Bridge up to Maidenhead, a length of 52 miles of open navigation, but in the 63 miles between Oxford and Maidenhead (with a fall of 108 feet) Bishop names 23 locks, of which 19 are known to have been in mill weirs and four in weirs built solely for navigation purposes.

These were all flash locks of the type with retractable paddles (and overfall boards) and rimers bearing on a pivoted swinging beam and a fixed sill (Figure 2.1). Such locks

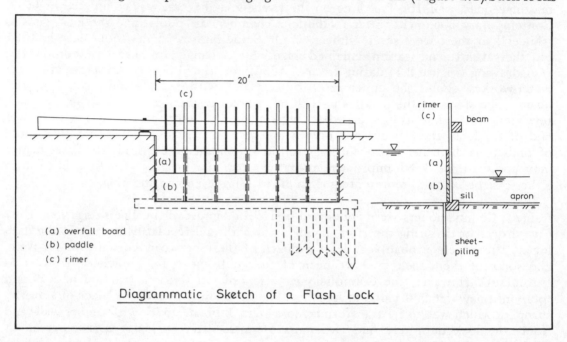

Diagrammatic   Sketch   of   a   Flash   Lock

Figure 2.1.

are of medieval origin and, after countless renewals, remained in use until the 18th century. A map of the river near Abingdon, dating from *c.* 1570, shows 'The Locke' with five sets of paddles.[6] Plot, in 1677, gives a clear description of the Thames flash locks,[7] and practically every one in Bishop's list can readily be identified in early 18th-century navigation rolls; while a well known sketch plan of 1786 shows the weir at Whitchurch with its sluice gate and a barge being winched upstream through the opened flash lock.[8]

The weirs themselves must have been considerable structures; built obliquely across the river, upwards of 400 feet in length, and impounding as much as six or seven feet of water with falls of four or five feet as a maximum differential head.

Seventeenth century

Thames: above Oxford   Commissioners appointed under an Act of 1605 succeeded in extending the navigation 'for many miles beyond the city of Oxford westward'.

These words date from 1624. More explicitly in 1695 the river is said to have been navigable 'for divers years last past . . . somewhat farther than Lechlade'.[9] No contemporary details of the work are available, but paddle and rimer flash locks were still in use throughout this part of the Thames until the late 18th century (indeed, a few remained until quite recent times), and there is no reason to suppose they differed in any essential way, in type or location, from the original structures. When toll lists and descriptions become available[10] they show barges of 40 to 50 tons burden travelling between Lechlade and Oxford through 20 flash locks in this length of 32 miles, in which the fall amounts to 50 feet. The locks, of the paddle and rimer type about 14 feet wide, were in timber-framed weirs up to 40 feet in total length.

Great Ouse (Part I)   The first integrated scheme of pound lock navigation on an English river was carried out by Arnold Spencer (1587–1655) on the Great Ouse between St Ives and St Neots.[11] In 1618 he and a partner acquired the right to undertake the work at their own charge, and on payment of an annual rent to the Crown, in return for any profits. In contemporary terminology Spencer built six 'sluices', but these were certainly pound locks, each with two pairs of mitre gates and a chamber large enough only for a single boat.[12] The chamber walls may have been in timber. The fall from St Neots to St Ives is 28 feet in a length of 15 miles,[13] and the locks were located in cuts or side streams at each of the mills in this part of the river. The work had been completed by 1625.

Spencer hoped to extend the navigation to Bedford. Soon after 1628 he built another lock at Eaton Socon (two miles above the St Neots lock), this time with a double size chamber; he also made two cuts about 100 and 200 yards in length and scoured the river with a horse-drawn scarifier. In this way he enabled barges to reach Great Barford, but financial difficulties prevented any further works.

Thames: Oxford–Burcot   In the last 15 miles up to Oxford the Thames has its steepest gradient (24 inches per mile) and there were many shoals; so the larger class of barge used Burcot as the head of navigation. To remedy this state of affairs Commissioners were appointed under an Act of 1624. Considerable sums appear to have been spent during the next five or six years, perhaps on dredging and general improvements, but the main achievement was the construction of three pound locks; at Iffley and at Sandford, replacing flash locks, and in a cut at the head of a re-opened channel of the river (the Swift Ditch) near Abingdon.[14] The locks were completed by 1638. Nuneham flash lock remained, to be used as a stanch for providing, when necessary, an adequate depth of water on the lower sill of Sandford lock. Barges calling at Abingdon continued to use the flash lock there. Interestingly, this was rebuilt in 1649 with excellent masonry abutments. The first pound lock at Abingdon, on the site of the present one, was not built until 1790.

The Oxford–Burcot pound locks, confusingly called 'turnpikes' and as such described by Plot in 1677, had masonry walls and two pairs of mitre gates. The chambers were about 80 feet in length and 19 feet wide, with 4 to 4½ feet of water on the sills. Sandford lock had a fall of 6 feet, which helps to explain why pound locks were needed, quite apart from their greater convenience and economy in water.

Warwickshire Avon   This river was first made navigable from Tewkesbury to Stratford during the years 1636–39 under the direction of William Sandys (1607–1669). According to an account by Thomas Habington (who died in 1647), Sandys built 13 'sluices' (at named sites, most if not all at mills) and 'erected also locks and placed many wires [weirs] in the quickest streams' and, seemingly a little later, relieved 'the grievous [*sic*] oppression of waters' on neighbouring land, 'by opening of two new floodgates'.

For the next owner of the navigation various improvements were carried out around 1650 on the river below Evesham, and the 17th-century work reached completion about 1665 with the construction or re-construction of five 'sluices'. These were built by Andrew Yarranton (1616–*c*.1684) acting as engineer for, and partner in, a syndicate holding a lease on the upper river.[15]

It is difficult to say exactly what was done in each of the three stages of construction, but the final result is known: a navigation 43 miles in length, suitable for 30-ton barges, with seven pound locks and four flash locks or stanches in the upper district (18 miles, 48 ft. fall) and seven pound locks and two stanches in the lower district (25 miles, 36 ft. fall).

Until recently four of the locks (at Pershore, Wyre, Chadbury and Cleeve) retained what was probably their original diamond-shaped chambers: the idea perhaps being to minimise erosion of the earth slopes by the rush of water from the upper gate paddles. The stanches situated below Fladbury and Pershore mills were classic early examples: normally left open so as not to impair the efficiency of the mill wheels, and closed only on the approach of a boat to raise temporarily the upstream water level for navigation before being used as a flash lock. The fall at these stanches, when closed, was two to three feet.[16]

Wey   The Wey navigation, designed by Sir Richard Weston (1591–1652), is a remarkable work of river engineering in the 17th century.[17] The Act dates from June 1651. Construction began in August and was carried on with great determination, clearly in accordance with carefully prepared plans. As many as 200 men were employed. Sir Richard found part of the capital, supplied most of the timber, and directed operations until shortly before his death in May 1652.

Work continued under George Weston, who had been instructed by his father, and the navigation was opened in November 1653. It had (and still has) a length of 15 miles, of which nine miles were in cut, with 10 pound locks, four new weirs, 12 bridges, and a wharf at Guildford (Figure 2.2). The total cost exceeded £15,000. All the locks were probably turf-sided and timber-framed, with falls ranging from four feet to eight feet, and together they accounted for rather more than 60 feet out of the total fall of about 72 feet from Guildford to the Thames at Weybridge. Flood gates were placed at the head of the two longest cuts, but they may have been later additions.

The navigation proved to be successful and seems scarcely to have needed any changes or major improvements for many years. In the 18th and early 19th century the cuts were shown as having a top width of about 52 feet[18] and the locks as being large enough to admit barges 84 feet long by 14 ft. 3 ins. wide.[19] These dimensions may well apply to the original construction.

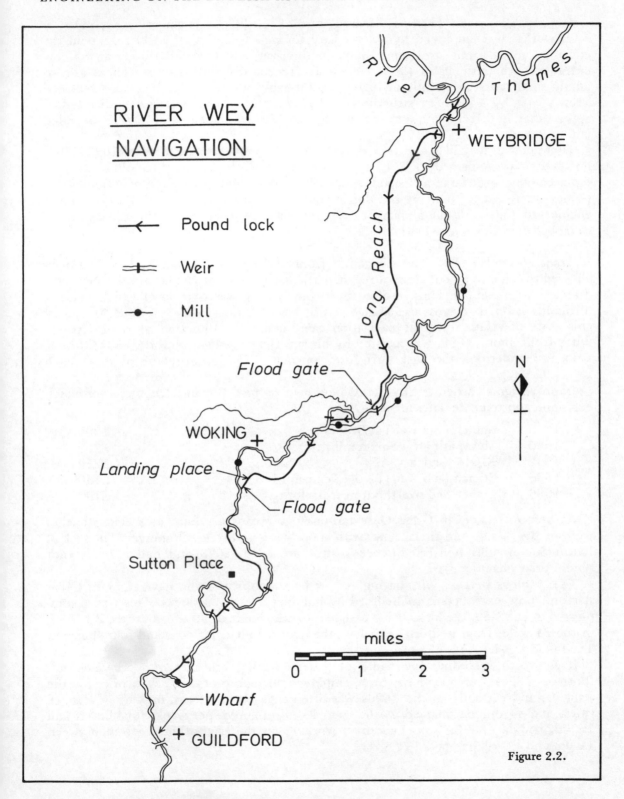

RIVER WEY NAVIGATION

←— Pound lock

╪— Weir

•— Mill

Flood gate

WOKING

Landing place

Flood gate

Sutton Place

N

WEYBRIDGE

River Thames

Long Reach

miles

0   1   2   3

Wharf

GUILDFORD

Figure 2.2.

Exeter Canal   In 1675–76 Richard Hurd of Cardiff, an engineer 'experienced in such works' was employed by Exeter City Council for a fee of £100 to extend the canal by about a mile to deeper water in the river, and to make a basin at the new entrance with mitre gates (later known as Trenchard's Sluice) of sufficient size to admit small coasting vessels at high tide. Probably at this time Hurd also replaced Trew's weir by a masonry structure nearby: some 250 feet in length with a fall of approximately 6 feet. Total expenditure by the City was upwards of £5,000 including land purchase.

The canal itself still could take only 16-ton lighters. During the years 1698–1701, however, it was enlarged to 50-ft. top width and 10-ft. depth; the three old locks were replaced by a single large turf-sided lock, with masonry abutments, and new flood gates (Kings Arms Sluice) built at the upper end of the canal. From 1699 Daniel Dunnel engineered the works at a salary of £3 per week. Navigation on the reconstructed canal began in November 1701.[20]

Great Ouse (Part II)   The navigation from St Ives up to Great Barford had been allowed to fall into a poor state of repair, particularly above St Neots, and Vermuyden's drainage works in the Fens resulted in a somewhat lower river level which made it difficult for boats to cross the shallows at St Ives. Nothing much was done to remedy this state of affairs until the navigation came under the direction of Henry Ashley who, from about 1680, was assisted by his son Henry Ashley junior (1654–c. 1730), who also undertook the Lark navigation, opened c. 1716, and rebuilt St Ives stanch in 1720.

Improvements carried out by the Ashleys involve the following structures, for which costs and approximate dates are known:[21]

1678	Stanch (and weir) at St Ives	£200
1680	Sluice at Eaton Socon mills	£230
1682	Roxton sluice	£250
1686	Stanch (and weir) below Tempsford	£130
1686	Stanch and overshoot near Barford	£100

As before, 'sluice' in Great Ouse terminology means a pound lock. The stanches were of the paddle and rimer type, with sluice gates and weir adjoining.[22] The lock at Eaton Socon mills replaced Spencer's structure of c. 1628 and had a single (not double) size chamber.

The Ashleys *père et fils*, having restored and improved the navigation to Great Barford, now extended it to Bedford by building five pound locks. This was accomplished in 1687–89, the locks being designed to take boats of 30 tons burden. By 1689, in the 31 miles from Bedford to St Ives, the river then had 13 locks and four stanches. The fall in this length is 65 feet.

It is worth recording the name of Joseph Hunt, carpenter of Godmanchester (born c. 1650). He may have been employed on some of the new works for the Ashleys, and certainly in the 1690s was often engaged in repairs, making new sluice gates, and so on. He charged 2s. 6d. per day, or 15s. 0d. per week, for himself and 1s. 0d. per day for his son. Labourers working on the navigation received between 1s. 0d. and 1s. 4d. per day.

Early eighteenth century

Aire and Calder   In the 17th century boats of 30 tons burden could go up the river Aire on the tide to Knottingley, 24 miles from the junction with the Yorkshire Ouse. Under an Act of 1699 the navigation was extended 23 miles to Leeds and, on the tributary river Calder, 13 miles to Wakefield. Improvements were also made in 12 miles of the river below Knottingley, to Weeland.[23]

The proprietors, named in the Act, consisted of nine members of Leeds Corporation, including the mayor, and nine gentlemen of Wakefield. The Leeds committee had employed the engineer John Hadley to survey the Aire in December 1697. He considered the project to be eminently feasible, requiring seven pound locks between Knottingley and Leeds, one at each of the mills, and deepening of the river in various places as far downstream as Haddlesey where, in a later opinion, a further lock should be built. At about the same time the Wakefield committee commissioned a survey of the Calder from Samuel Shelton. Also, at the request of Parliament, two members of Trinity House studied the possible effect the works might have on the tidal regime of the Ouse and on the trade at various riverside towns. Their sensible and impartial report was written in July 1698.[24]

Hadley gave evidence during the Parliamentary proceedings in 1698 and 1699, and on 9 May 1699 was appointed engineer for the river Aire at a fee of £420. Work began without delay and, despite some damage by a flood in January 1700, the navigation was opened to Leeds in November of that year, the seven locks above Knottingley having been completed as well as a warehouse at Leeds. Not long afterwards two additional locks were built on the Leeds line, one at Methley and the other at the tail of the 1½-mile Cryer Cut newly excavated to bypass a very circuitous stretch of river near Woodlesford.

It is probable, though not yet confirmed, that Hadley died in 1701 (before August of that year). Thereafter the proprietors seem to have relied on James Mitchell, as engineer-contractor, for technical aspects of the works. A carpenter from Bradley, near Huddersfield, he had been involved in building some of the Aire locks. In July 1701, in partnership with James Willans, mason of Dewsbury, he contracted to make the Calder navigable. Four locks were built between Castleford and Wakefield, and this part of the navigation was opened in the summer of 1702.

Meanwhile it had been decided to have two locks (not one) below Knottingley: at Beal and Haddlesey. In April 1702 Mitchell and George Atkinson (1663–1728) of Thorne signed a contract to build these locks and their weirs, and by November 1704 it could be said that the whole navigation 'was near perfected'.

Of the 15 locks, eight were in cuts at the mills, as illustrated in Figure 2.4,[25] but the others required new weirs ('dams' in north country terminology) across the river. As a result of experience, three additional sluice gates had to be provided in Beal dam to avoid flooding. There were flood gates at the head of Cryer Cut; in places the river embankments had to be raised and doubtless a good deal of dredging was needed.

The locks had masonry walls and timber flooring, and were about 60 feet long by 15 feet wide with 3 ft. 6 ins. of water over the sills. At the outset boats appear to have been limited to 15 tons burden but subsequent improvements in the river, without alteration to the locks, enabled boats to carry 28 tons on average; more in

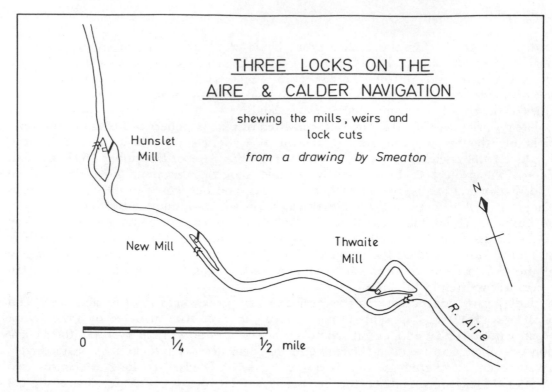

Figure 2.3.

winter, less in a dry summer.[26] The whole undertaking cost over £25,000 but proved to be useful and profitable.

Essex Stour   The Corporation of Sudbury obtained an Act in 1706 to make the river Stour navigable from their town to Manningtree at the head of the estuary, a length of 25 miles. Shares in the undertaking were mostly purchased by two London merchants who took responsibility for carrying out the works. The navigation was opened throughout in May 1709, at a total cost of around £9,000. It had 13 pound locks, timber-framed and turf-sided, and 13 stanches of the paddle and rimer type.[27]

Later in the 18th century the locks were walled in timber but retained their characteristic overhead lintels between the gate posts used, instead of land ties, to resist the earth pressure. These and other features of the Stour are well known from the paintings of John Constable, whose father was one of the navigation commissioners from 1781.

Kennet   From Newbury to the Thames at Reading the Kennet has a gradient of 66 inches per mile and a correspondingly swift current with many shoals. Under an Act of 1715 work began on building locks at the mills along the river, but it became apparent that success was not going to be achieved in this way. In the hope of finding a solution the proprietors in 1718 appointed John Hore (*c.* 1690–1763) as their engineer. His plan involved cuts totalling 11 miles in length and 20 pound locks. Completed in 1723, the navigation had a length of 18½ miles with a wharf and barge basin at Newbury, wharfs at Aldermaston and Reading, and several new bridges.[28].

From Reading to Sulhamstead there were seven cuts, up to three-quarters of a mile long, bypassing four mills and three bad sections of river where weirs had to be built. Further upstream the navigation was predominantly in cut except for about a mile near Woolhampton and another mile or so near Thatcham. (Figure 2.4). Here some dredging and embanking were needed and a run-off or waste weir.

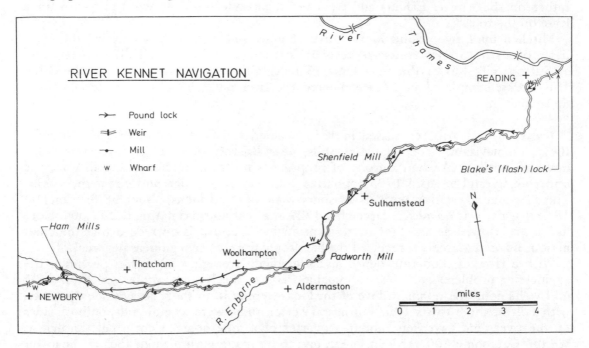

Figure 2.4.

The two locks above and below Aldermaston wharf were in brick, and so (probably) was the lock at Reading. All the rest were timber-framed and turf-sided. Evidence given in 1725 that 100-ton barges passed fully laden from Newbury through Reading to London suggests that from the start the locks had dimensions not very different from 114 feet by 17 feet as in three of the surviving turf-sided locks. Falls ranged typically between five and nine feet.

A letter written by Hore in July 1721 reports progress. Two of the longer cuts are now opened he says, two locks near Reading have recently been framed, the banks near Thatcham are being formed using 'the two large boats' (presumably dredger barges) and work is starting on Padworth cut with its two brick locks. He also refers to the Archimedean screw used in draining the excavations.

According to usual practice at this period, several small local contractors would have been employed, carpenters, bricklayers, blacksmiths for the iron work, and so on. Some of their surnames by chance are known: Potenger of Sulhamstead and

Lawrence of Burghfield, labourers (i.e. engaged on excavation), with Hughes of Shenfield, Grace of Midgham, and Saxon and Slade of Newbury, carpenters.

From an engineering point of view the Kennet is the outstanding early 18th-century navigation. The total cost was upwards of £45,000.[29]

Yorkshire Derwent   The engineer George Sorocold surveyed the Derwent in 1694 and again in 1699. The Act for making the river navigable was passed in 1701 but little, if anything, was done until 1720. Joshua Mitchell (who had worked with his father on the Aire & Calder) and a partner then undertook the work in return for a lease on the tolls.[30]

Mitchell built five pound locks, three at mills and two with new weirs; the locks being of a size to take Yorkshire keels 55 feet long by 14 feet beam. The lowest lock, at Sutton mill, was at the tidal limit 15 miles above the confluence with the Ouse. The works, completed in 1724, rendered the river navigable for a further 23 miles up to Malton.

Bristol Avon   An Act passed in 1712 enabled the Corporation of Bath to make the Avon navigable from Hanham Mills, near Bristol, to Bath. Nothing was done, however, until 1724 when a group of proprietors took financial responsibility for the undertaking, and in April 1725 appointed John Hore to design and direct the works. The 11½-mile navigation with six masonry-walled pound locks, about 98 feet long by 18 feet wide, was opened in December 1727 at a cost approximating to £12,000. Hore had a Mr. Downs as assistant or resident engineer, and it is on record that the chief mason, Edward Marchant, received the handsome wage of two guineas per week.[31]

With a river of moderate depth and a gradient averaging 30 inches per mile, the engineering problems were less formidable than on the Kennet; moreover the existing mill weirs could be retained. Five of the locks were built in short cuts beside the weirs, with falls of three to six feet. But near Twerton there were several mills, on both sides of the river, and two weirs about a quarter of a mile apart. Here a cut, known as Weston Cut, some 600 yards in length had to be made with a single lock at the lower end accommodating the total fall (about nine feet) of the two weirs combined.

A large-scale map of 1742 by Thomas Thorpe shows this cut and the next three locks downstream at Kelston, Saltford and Swinford, with the weirs and mills.[32] It is interesting to note that the weirs, like those on the Aire (Figure 2.3), are oblique.

Mersey & Irwell   The plan by Thomas Steers (*c.* 1672–1750) of the Mersey & Irwell, dating from 1712, is one of the earliest published for any river. Levels are given at nine points along the rivers from Manchester to Warrington (fall 52 feet) and the scheme involves eight pound locks and two cuts.[33]

The locks were built between 1724 and 1736, and also a wharf at Warrington, but whether Steers acted as engineer is not certain. The cuts, to eliminate long loops in the Mersey, followed later.

Don   In favourable conditions the river Don was navigable, on the tide, for about 16 miles above Goole to Wilsick House and occasionally for a further eight miles to Doncaster. As part of an effort to improve the navigation Doncaster Corporation sought advice from Joseph Atkinson (1693–1760; son of George Atkinson, the Aire & Calder contractor) and, at the same time, the Cutlers' Company employed William Palmer (d. 1737), engineer of York, to plan an extension up to, or near, Sheffield.[34] Working jointly, with help from Joshua Mitchell, they produced a map of the river in October 1722, to a scale of 2½ inches to a mile. Next year this was published as an engraving (on a reduced scale) with tables of lengths and falls.[35]

Above Doncaster the river was rather steep and swift flowing with dams (weirs) for the mills at places such as Kilnhurst, Rotherham and Sheffield itself. The solution here clearly involved cuts, some of considerable length, with locks, but making use of the existing weirs to avoid further impeding the flood flow. The last four or five miles up to Sheffield were evidently going to be difficult, however, and it was soon decided to terminate the navigation at Tinsley. On this basis an Act passed in 1726, with Palmer giving evidence.

Below Doncaster the river had a very different character, flowing through low-lying country with a gradient of scarcely more than 20 inches per mile from the tail of Doncaster mill dam to Wilsick, and in particular with no weirs. Nevertheless this part of the river had to cope with floods coming down from the higher ground upstream, and it also acted as the outfall for several land drainage schemes in the neighbourhood. Consequently the landowners opposed any measures which might lead to flood or other damage to their property.

At first the plan was to scour the river and make cuts, without locks or weirs, past the worst of the shoals. Then came the idea of building an opening weir with a lock at one especially bad place, and finally it was agreed to have locks and weirs not more than four feet high with very ample sluice gates. Atkinson explained the scheme to Parliament in proceedings leading to the Act of 1727, his evidence being confirmed by Richard Ellison (c. 1686–1743) of Thorne who on this and later occasions appeared as expert witness for the Don.

Though it could be assumed that Palmer directed operations for the Sheffield proprietors, the early history of the works is obscure. By 1731, however, about £12,000 had been spent and the navigation now extended seven miles above Doncaster. The two groups of proprietors then amalgamated, formed a management committee and in August 1731 appointed John Smith of Attercliffe as engineer. Described as carpenter or millwright, terms virtually synonymous in a river context, he held this post until 1767.

By 1733 ten locks had been built, eight from Doncaster upwards and two below. A third, at Barnby Dun near Wilsick, followed a little later. The original plan, with a few variations, was now realised except for the last five miles up to Tinsley, and included a 2½-mile cut past Kilnhurst with three pound locks and flood gates at the head. The locks throughout were of the same length as those on the Aire & Calder and one foot wider.

The need now became apparent to bring the river below Wilsick to the standard achieved upstream. Accordingly an Act of 1740 gave powers to make a 2-mile cut, with two locks, between Bramwith and Stainforth and to deepen the channel down to Fishlake. At the same time the plan for completing the navigation to Tinsley was revised. All these works were finished by 1751, resulting in a navigation 33 miles long with nine miles of cut and 17 locks (and two flood gates) suitable for 30-ton vessels.

Smeaton, writing in 1767, said he had seen several of Mr. Smith's locks 'which have stood and answered very well': praise indeed from that master of design and construction.[36]

Yorkshire Ouse   In February 1699 the Corporation of York requested the members of Parliament for the city 'to send an Engineer as soone as possibly they can to view the River of Ouze in order to make it more navigable'. They made a good choice, and on 4 May Thomas Surbey arrived from London. (Nothing else is known of this evidently able engineer; his MS report on the Ouse is in York City Archives.) His report, dated 22 May, is accompanied by a chart with soundings and plans of the proposed works, which include a dam and lock at Naburn (six miles below York) and a 1,200-yard cut near Booth Ferry.[37] The Corporation paid Surbey £24 plus expenses, and sought a second opinion from John Hadley. He agreed with the idea of Naburn lock.

Thus encouraged, York sent a petition to Parliament for leave to bring in a Bill. This faltered at the second reading, however, and little more was done until 1725 when William Palmer received instructions to prepare a survey. Part of his map, which includes tidal data, is shown in Figure 2.6.[38] The city now succeeded in obtaining an Act, passed in 1727, to levy navigation tolls to meet the cost of improvement works.

The precise nature of these seems still have have been in doubt. The Naburn project though attractive would be expensive; John Perry (c. 1669–1733), asked for his opinion, recommended a flash lock above Cawood and several bypass cuts totalling 2½ miles in length, while Palmer favoured a mile-long cut near to the mouth of the Wharfe. In all cases, however, some deepening of the channel would additionally be required.

On grounds of cost the latter expedient was adopted as the sole means of improvement. Palmer directed the operations from 1728, on a retaining fee of £40 p.a. with payments for special works. Timber jetties or training walls were built to contract the stream, and so increase its velocity on the ebb, and the river bed was scoured or 'ploughed'.

Work along these lines continued for several years and doubtless proved effective in removing deposits of tidal silt. But at several places with notorious shoals the bed consists of boulder clay or clayey gravel which would have proved too resistant to respond to this treatment or, indeed, to the 'vigorous dredging' (with, of course, hand-operated equipment) proposed in 1748 by John Smith, who had been brought from the Don to give his advice on the problem. And so at last in 1752 came the decision to build Naburn lock and dam. An account of these works is given later.

Weaver   The river Weaver was made navigable from Frodsham up to Winsford, a distance of 18 miles, by the construction of 11 pound locks and weirs.[39] Work began in 1730 under the direction of Thomas Robinson as 'surveyor general'. The locks were finished by the end of 1732 and the whole job completed in 1735, at a cost of roughly £18,000.

An unusual feature is that the locks were built within the river (not in cuts) in locally widened sections to give an adequate weir length. The entire structure of lock, sluice and weir was in timber. The locks, about 60 feet long and 16 feet wide, could accommodate boats up to 50 tons burden.

Lee An engraved map of the Lee navigation in 1740, with levels along the river from Hertford to Bow, shows a fall of 111 feet in this distance of 31 miles.[40] The mill at Ware is situated on the river bank with a 'cistern' (pound) lock nearby, but the 11 mills further downstream are on rather long cuts or side streams. Across the river below the mill-stream intakes, and at some other places, there are four 'turnpikes' and 15 'weirs'. All of these are undoubtedly flash locks. No description of the 'weirs' has been found; probably they were of the paddle and rimer type. But it is certain that 'turnpikes' were guillotine gates with an overhead windlass of the pattern familiar in East Anglia (Plate 3).

In 1741 a new turnpike was built near Broxbourne to the design of William Whittenbury (d. 1757), resident engineer to the navigation, whose map has just been mentioned. The main points in his specification are: the gate to be 14 feet wide and fitted with friction wheels to 'ensure its motion' by the 'Great Wheel' and windlass; and beside the lock gate there are to be three sluices each 6 ft. 8 ins. wide, making a total opening width of 34 feet. When all the gates are closed, a depth of 4½ ft. of water will be impounded above sill level, with a fall of 3 ft. 3 ins. Plank aprons to extend 12 feet upstream and downstream, to protect the bed from erosion.[41]

It is known, in general terms, that a large quantity of water is 'lost' in the operation of flash locks on a busy navigation. But Smeaton and Yeoman, in 1767, actually measured (or, rather, made a careful calculation of) the average discharge in a dry season through a Lee turnpike at Waltham Abbey and found the astonishing figure of two million cubic feet per day.[42] This is equivalent to at least 250 lockfuls in a typical pound lock. No wonder there were frequent conflicts between millers and boatmen, and one is surprised at the long delay in replacing flash locks by pound locks on the Lee and the Thames, an improvement for both rivers which came on a serious scale only around 1770.

Denver Sluice This important work of river engineering, as completed in 1750 across the Great Ouse, is shown in an early 19th-century drawing reproduced in Figure 2.8.[43] High embankments run alongside the river downstream to King's Lynn; the sluice prevents the tides from overtopping the lower banks upstream on the Great Ouse and other South Level rivers, and the lock provides for navigation at all normal states of the tide. There are in effect two separate structures: (i) The main sluice built in stone-faced brickwork with three arches each giving an 18-ft. clear waterway and equipped with a pair of self-acting pointing gates (the 'sea doors' of Plate 5) and another pair pointing upstream, used in dry seasons to maintain an adequate depth for navigation to Ely and beyond, and for agricultural reasons. The sea doors rise to the level of a large spring tide (1.5 feet above H.W.O.S.T.) which is 17.75 feet above sill level and 19.25 feet above the apron. (ii) A two-arch brick structure on the east side of a 'lock island' containing a supplementary 16-ft. sluice and a navigation lock 52 feet in length between gates, 13 ft. 3 in. clear width, and with 5 ft. 3 in. depth on the sills at normal low water. Like the sluices, the lock has a pair of ebb gates or 'land doors' for dry seasons.

The principal materials used were: locally-made bricks, stone from Ketton, tarras and lime mortar, Baltic fir for bearing piles, sheet piles and planked floors, and oak for the gates. As well as the usual barrows, planks, pumps, etc. there was a winch-operated pile driver and a machine for dovetailing the sheet piles. Work began in June 1746, the lock was opened exactly a year later, the main sluice came into operation in November 1749 and the whole job reached completion by the end of 1750. The costs were approximately: lock £1,100, cofferdams £900, main sluice £3,100.

The work was directed throughout by William Cole, Surveyor (i.e. Engineer) of the South Level from 1731 to 1750, with John Leaford (Divisional Officer 1729-53) as resident engineer. William Robinson of Ely contracted for the lock and cofferdams but the main sluice was built by direct labour. Besides the minute books of Bedford Level Corporation, the authority then responsible for all river and main drainage works in the Fens, surviving original documents include Cole's journals for 1746 and 1749, his meticulous accounts for the entire period of construction, a plan of the scheme in 1747 by Leaford, and a report dated June 1748 by Charles Labelye (1705-1762) who was consulted on the main sluice foundations.[44]

Denver Sluice has a long history going back to the first works by Vermuyden in the 1650s. Here it is relevant to note that the main sluice of 1750 replaced a similar but smaller structure built 1682-84 which suffered a foundation failure in 1713, and had to be demolished; and that the two eastern brick arches were originally built in 1699 by Richard Russell, a predecessor of Cole, for two additional sluices.[45] The lock and supplementary sluice made use of Russell's pier and side walls. A report by Labelye in 1745 ended a strangely long period of indecision as to whether, and how, Denver Sluice should be rebuilt but, although he is usually credited with the design, the 1750 structure differs very materially from his proposals.[46] No changes occurred thereafter until the 1830s when, to derive full benefit from the deepened river bed brought about by the Eau Brink Cut, the present lock and sluice were constructed to designs by Sir John Rennie.[47]

Naburn Lock and Dam   Built on the Ouse six miles below York, Naburn lock and dam are the most impressive single works of their kind on an English river before 1760. The contract, dated 18 September 1752, is signed by John Smith of Attercliffe and his son John Smith junior.[48] By February 1753 agreement had been reached with the local landowner on rent payable for the ground to be cut and otherwise to be used by the navigation,[49] and construction started later that year or early in 1754.

John Smith junior (c. 1725-1783) acted as site engineer, payments under the contract being made to him,[50] but it can be taken that his father was responsible for design and expert supervision. Work proceeded rather slowly, due to the limited funds available as well as to the engineering difficulties on a river subject to large floods, and was completed in 1757 at a cost not exceeding the contract price of £5,500.

A site plan and a section through the lock chamber are shown in Figures 2.5 and 2.6. The leading dimensions and a brief description may be summarised as follows:[51]

> Lock Cut: About 350 yards in length, top width 54 feet, excavated 16 feet below ground level and embanked on the east side.
> Lock: 90 feet between gates, 140 feet overall, 20 feet clear width, 23 feet deep from coping to the plank floor. Masonry walls, slightly battered on the front face and backed by counterforts to increase resistance against lateral earth and ground water pressure, founded on timber beams 43 feet long and about 16 inches deep. The wing walls founded on timber piles. Lower sill 22 feet and upper sill 18 feet below coping, these dimensions also being approximately the height of the gates.

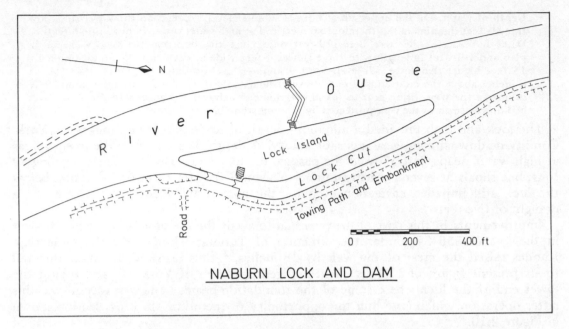

Figure 2.5.

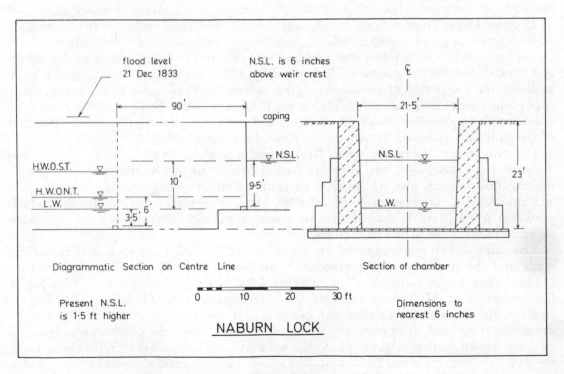

Figure 2.6.

Depth of water: On the upper sill at normal summer level 9½ feet; on the lower sill at low tide 3½ feet (maximum fall therefore 10 feet) and at high water ordinary neap tide 6 feet.
Dam: Built across the river, here 170 feet wide, consisting of a masonry weir V-shaped in plan and 140 feet in length with three sluices 4 feet wide at each end. Weir crest originally 13 feet higher than lower sill level. Normal summer level regulated by the sluices at about 6 inches above the crest and 10 feet above ordinary low tide. Floods in winter completely submerge the weir, rising near to (or in extreme cases above) coping level of the lock. River bed downstream of the weir protected by random stone blocks.

The lock and dam provided a minimum depth of seven feet at all times up to York. Conditions downstream were less satisfactory. A depth of six feet on the lower sill, as at high water neaps, just allowed the passage of fully laden Humber keels carrying 80 tons, but shoals at several places (notably Acaster Selby) for three or four miles below the lock still impeded navigation even for the more usual loading of 60 tons on a draught of five feet.

Improvements in this part of the river had to await the use of a 10-hp steam dredger in the years 1835–37, under the direction of Thomas Rhodes. At the same time Rhodes raised the crest of the weir by 18 inches,[52] thus bringing the maximum fall to its present figure of 11½ feet, and dredged upstream of York. He also rebuilt the lower end of the lock where some of the foundation beams had given way. It was this latter operation which gave him the opportunity of recording the cross section shown in Figure 2.10.

Nene Navigation was always possible on the tidal river from Wisbech to Peterborough, or perhaps a few miles beyond. Commissioners appointed under an Act of 1725 agreed with Thomas Squires and a partner that he would make the river navigable from Peterborough to Oundle, with the right thereafter to levy tolls, and this was done in the years 1726–30. Squires then became sole proprietor and, under a similar agreement, extended the navigation to Thrapston by 1737. No minute books appear to be available for the period of construction, but references in the later 18th century show that pound locks (still called 'sluices' as on the Great Ouse) had been built at each of the 14 mills in this 35-mile length of river, and probably about half a dozen stanches of the guillotine type well known on the Nene until recent times.

The river from Thrapston up to Northampton was surveyed in 1744 by Thomas Yeoman (*c.* 1708–1781) but 10 years passed before an active interest developed in making this western part of the river navigable. Yeoman went over the ground again and produced a rough estimate of £12,000. He also gave evidence on the Bill, which passed in April 1756. Commissioners were appointed and within a year the necessary capital had been raised by subscription.

Yeoman's sketch map indicated on which side of the mills the lock cuts should be made and the positions of the stanches.[53] In November 1757 Ferdinando Stratford (1719–1766) began more detailed surveys. John Case of London and John Smith (accompanied by his father) attended the Commissioners in March and April respectively with plans and proposals for carrying out the work. Yeoman and Stratford examined these and, after considering several other tenders, the contract was awarded to John Smith junior in June 1758. His estimate, with detailed specifications, came to £14,070 for building 20 pound locks and various ancillary works including a total length of 3¾ miles of cut; the locks to be 10½ feet wide at the bottom and

100 feet long between the gates (to take two barges in tandem) with battered stone-faced brickwork walls and timber floors.[54]

Smith acted as engineer and contractor: Yeoman continued in a consultative capacity only, and a clerk was appointed to certify payments for work done. Smith appears to have handled the job efficiently. He made some variations in detail and was asked (and paid) to carry out a few additional items. Yeoman inspected the works in October 1760 and again in January 1762, on the latter occasion signifying his complete approval of the navigation, the formal opening of which had taken place in August 1761. The total expenditure amounted to £15,650.

The cuts by-passed 21 mills and were carefully planned to improve the course of the river. In addition to the 20 locks (one of which, on a long cut, accommodated the fall at two mills) there were two or perhaps three stanches, and a minimum depth of four feet obtained in all parts of the river. A horse towing path, with some 20 crossing bridges, had been provided along the whole length of 26 miles and wharfs were built at Wellingborough and Northampton.

The total fall of 180 feet from Northampton to Peterborough was accounted for approximately as follows: 136 feet in the locks, 10 feet at the stanches (when closed) and 34 feet in the river.[55] The profile in Figure 2.7 shows the division of the fall between locks and river in a typical length of the western navigation.

Our story ends in the years around 1760 with the beginning of the Canal Age and the emergence of a civil engineering profession under the leadership of John Smeaton. In the preceding period the river engineers were as important as those engaged on harbour works, fen drainage and bridge construction, and left as their memorial many hundreds of miles of navigable inland waterways together with much experience of benefit to the next generation.

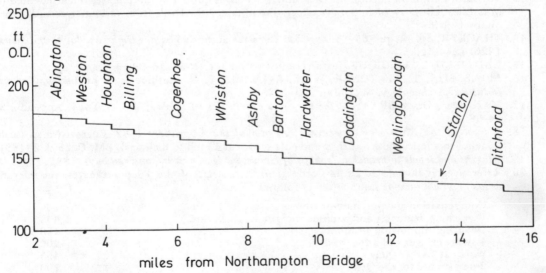

Profile of Part of the Nene Navigation

Figure 2.7.

NOTES AND REFERENCES

1. Memoir of the Canal of Exeter, from 1563 to 1724. By P. C. De la Garde; with a Continuation by James Green. *Min. Proc. Inst. C.E.* 4. (1845). 90–110. STEPHENS, W. B. The Exeter lighter canal. *Journ. Transport History.* 3. (1953). 1–11. CLARK, E. A. G. *The ports of the Exe Estuary.* 1960. 28–40.

2. Bodleian Library: Twyne-Langbaine MS, I, f. 76. The drawings and notes are by Brian Twyne of Oxford University who visited Exeter Canal about 1630 when work was starting on the Thames locks at Iffley and Sandford. PRICHARD, M., & CARPENTER, H. *A Thames Companion.* 1981. 78–80.

3. FAIRCLOUGH, K. R. The Waltham pound lock. *History of Technology.* 4. (1979). 31–44. Some corrected dates have kindly been supplied by the author in a personal communication.

4. SALISBURY, C. R. Documentary evidence for an Elizabethan pound lock on the river Trent in Nottinghamshire. *Trans. Thoroton Soc.* 86. (1982). 115–117. Dr. Salisbury reproduces the relevant portion of the 1592 plan.

5. THACKER, F. S. *The Thames Highway; general history.* 1914. 54–55.

6. The original is at Abingdon in the Council Chamber. Photocopies are in BM: Maps 188. a (10) and RGS. S. 242.

7. PLOT, R. *The natural history of Oxford-Shire.* 1677. 232–233.

8. Reproduced in THACKER (*op. cit.*) frontispiece, and opposite p. 66 of HADFIELD, C., & SKEMPTON, A. W. *William Jessop, Engineer.* 1979. The sketch is by John Call, one of the Thames Commissioners.

9. Quotations from preamble to the 1624 and 1695 Acts for Thames navigation.

10. Toll lists for 1731 etc. in *Extracts from the Navigation Rolls of the River Thames and Isis.* 1772. Dimensions of the flash locks and adjacent weirs are given in the printed reports (1791 and 1802) by Robert Mylne on the upper Thames. It is unlikely that the locks had changed in any essentials since the 17th century.

11. WILLAN, T. S. *The navigation of the Great Ouse Between St Ives and Bedford in the seventeenth century.* Bedfordshire Historical Record Soc. Vol. 24, 1946.

12. SUMMERS, Dorothy. *The Great Ouse.* 1973. Plan opposite p. 53.

13. For information on falls of the Ouse and other rivers quoted in this chapter, I am indebted to Miss Elizabeth Shaw of Imperial College who obtained the levels from various river authorities on my behalf.

14. THACKER, *op. cit.* pp. 65–83. See also the same author's *Thames Highway: locks and weirs.* 1920. 125–152.

15. HADFIELD, C., & NORRIS, J. *Waterways to Stratford.* 1962. 15–24.

16. LEWIS, M. J. T., SLATCHER, W. N., & JARVIS, P. N. Flashlocks on English waterways. *Industrial Archaeology.* 6. (1969). 209–253.

17. MANNING, O., & BRAY, W. *History and antiquities of Surrey*, Vol. 3, 1814. Appendix 3, 54–58.

18. *A Plan of the River Wey. Surveyed by order of the Proprietors 1782.* Engraved map (scale 2 inches to a mile, widths of river and cuts shown at 1 inch to 8 chains). BM: C. 25 d. 2 (189).

19. *Lengths and levels to Bradshaw's maps of canals, navigable rivers, and railways.* 1832. 15.

20. References as in note 1. De la Garde gives a transcript of the 1699 estimates, the relevant items in which may be summarised as follows:

	£
Compensation and purchase of land	700
Deepening, widening and banking the canal.	1,462
Diverting Alphington Brook.	150
4 swing bridges and 2 footbridges	200
Works at Exeter Quay.	455
2 new cranes for the Quay	100
Kings Arm Sluice	800
New Lock	1,563
Repairing Trenchard's Sluice	55
Dredging the river	100
	£5,585

The estimates quote a sum of £605, additional to the figure given above, for walling the lock chamber but this was not carried out. The engineer engaged in 1698, William Baily, absconded in May 1699 before much work had been done. The appointment and salary of his successor Daniel Dunnel are noted by WILLAN, T. S. *River Navigation in England*. 1936. 81.

21. WILLAN, T. S. *Navigation of the Great Ouse. (op. cit.)*.
22. The paddle and rimer stanch at Tempsford and its mode of operation are described in PROVIS, W. A. On the locks commonly used for river and canal navigations. *Trans. Inst. C.E.* 1. (1836). 53–54.
23. UNWIN, R. W. The Aire and Calder Navigation. *Bradford Antiquary*. 42. (1964). 53–85; and 43. (1967). 151–186.
24. The report, by John Clements and John Bronwell, is transcribed on pp. 82–85 of Unwin's 1964 paper.
25. Smeaton's plan of the three locks on the Aire is in his collection of drawings in the Royal Society, vol. 6, fo. 18.
26. Smeaton, in a report dated 14 September 1767, printed in *Reports of the late John Smeaton*, vol. 1. 1812. 157.
27. CUBITT, W. (in *Min. Proc. Inst. C.E.* 4. (1845). 111–112) describes the stanches on the Stour as they were before improvements of the navigation in the 1830s.
28. THACKER, F. S. *Kennet Country*. 1932. An 18th-century MS plan of the navigation (scale 3 inches to a mile) shows the cuts, locks (with their falls) and bridges. Inst. Civ. Eng. library. Page Collection: f. 15.
29. *Journ. House of Commons*. 20. 288. 7 March 1726/7.
30. DUCKHAM, B. F. The Fitzwilliams and the navigation of the Yorkshire Derwent. *Northern History*. 2. (1967). 45–61.
31. CLEW, K. R. *The Kennet & Avon Canal*. 1968. 15–18, and extracts from the Minute Books of the Proprietors, kindly supplied by Mrs. Brenda Buchanan.
32. *An Actual Survey of the City of Bath, in the County of Somerset, and of Five Miles Round*. Survey'd by Tho: Thorpe in the Year 1742. Engraved by James Cole (scale 4 inches to a mile). BM: K.37 (9).
33. *A Map of the Rivers Mersey and Irwell from Bank-Key to Manchester with an Account of the rising of the Water and how many Locks it will require to make it Navigable*. Survey'd ... by Tho: Steers 1712. I. Senex sc. (scale 3 inches to 5 miles). Chetham's Library: Manchester Scrap-Book, item 16. Bodleian: Gough Maps 44 fo. 127.
34. WILLAN, T. S. *The early history of the Don Navigation*. 1965.
35. *A Survey of the River Dunn in order to improve the Navigation from Hull to Doncaster and to continue up to Sheffield*. By William Palmer & Partners taken Anno. 1722. E. Bowen sc. (scale 1 inch to 1½ miles). BM: 1145 (1). Bodleian: (E) c. 17 (451) [30] with accompanying report printed 1723.
36. Smeaton's report on Linton dam (June 1767) in *Reports*, vol. 1, 311.
37. DUCKHAM, B. F. *The Yorkshire Ouse*. 1967.
38. *A Map of the River Ouse ... Surveyed by Willm. Palmer & Partners in December 1725* (scale 1 inch to 2 miles). BM: K.6 (7). Bodleian: Gough Maps 41K fo. 32.
39. WILLAN. T. S. *The navigation of the River Weaver in the eighteenth century*. 1951.
40. *A Plan of the River Lee from Hartford to Bow Bridge with a Profile of the Fall* (by William Whittenbury, 1740). Scale 1.33 inches to a mile. BM; K.6 (6). Royal Society: Smeaton 6 fo. 87.
41. BOYES, J. *Canals of Eastern England*. 1977. 19.
42. Smeaton's report on the mills at Waltham Abbey, 16 May 1767. *Reports*, vol. 1. 279.
43. County Record Office Cambridge (CROC): R59/31/40/61(i).
44. Bedford Level Corporation records. CROC: R59/31/7 (Cole's journals), 10 (Conservators' Proceedings), 11 (Corporation Order Books), 19 (Account Books). The 1747 plan is R59/31/40/61(ii).
45. Dates, etc. of the late 17th-century structures are derived from R59/31/10, 11 and 19.
46. LABELYE, C. *The result of a view of the Great Level of the Fens*. 1745. His drawings, dated August 1745, are R59/31/40/61(iii) and (iv).

47. Rennie's sluices and lock were built in 1832–35 on the east side of the 1750 sluices with a sill level 6.0 feet lower. His drawings and reports are in the Institution of Civil Engineers library.

48. The original contract is in York City Archives: Acc. 203. I am indebted to Mr. M. F. Barbey for a summary of its contents.

49. Hull University Library: DDPA 7/721 and 722.

50. Accounts Book of the Ouse Navigation Trustees 1724–1834, also in York City Archives. Payments, probably for materials, begin in November 1752. Regular monthly or more frequent payments start in February 1754 and continue to November 1757, with a final settlement in 1758.

51. Details of Naburn lock and dam are chiefly taken from two reports by Thomas Rhodes dated 18 January 1834 and 13 December 1838, both reprinted with a map and section of the lock in *Reports and Plans of the River Ouse Navigation 1834 to 1881* (1881), and his lithographed *Longitudinal Section of the River Ouse from Linton to Selby* (1834) (scale 3 inches to a mile and 1 inch to 20 ft.) in Humberside Record Office: DDMW 7/390. Further details including tide levels and site plans are given in reports by John Coode (18 October 1876) and W. H. Bartholomew (6 May 1880), also reprinted in *Reports and Plans*. A second, larger lock was built alongside the original one in 1888.

52. RENTON, M. On the improvement of navigable rivers, with a description of a self-acting wasteboard at Naburn Lock. *Min. Proc. Inst. C.E.* 1. (1840). 26–27. Renton was Rhodes' assistant.

53. Yeoman's *Plan of the River Nen, from Thrapston to Northampton with the Mills and Locks necessary for the Navigation* (1754) is reproduced in HATLEY, V. A. *Northamptonshire past & present.* 6. (1981). 216.

54. This account of the Nene navigation is based on the Minutes of the Commissioners for the Western Division (1756 ff.) examined at Oundle by courtesy of Mr. G. E. Bowyer, Chief Engineer Welland and Nene River Authority.

55. From levels taken in 1931–32, kindly supplied by Mr. Bowyer.

CHAPTER 3

PUSH-TOWING – PAST AND PRESENT

WITH THE EXPLOSION of interest in Britain's canal heritage over the last two or three decades, the 70-ft. long, 25-ton capacity narrow boat, symbol of a bygone age in the development of inland transport, has acquired a romantic appeal equivalent to that of the steam locomotive for railway enthusiasts. It is doubtful whether a comparable romantic appeal will ever be associated with the subject of this chapter, which is the modern push-tow and the way in which it was developed. The push-tow has become the most efficient means of transporting bulk freight inland, and to give an idea of its remoteness from the British canal scene familiar to most readers, the carrying capacity of a single tow often amounts to a thousand times that of a narrow boat.

The push-tug itself (called a 'towboat' in America), with two, three or even four screws, packing the power to move up to 15,000 tonnes on the Rhine or 50,000 tonnes on the Mississippi, is an ungainly vessel, square on the bow, with two clumsy-looking uprights, and almost square on the stern. The uprights are towing knees, against which barges are securely lashed to form a rigid unit. This is the intrinsic beauty of push-towing. The motive power is separated from the carrying capacity, offering the same operating advantages as in conventional towing, but when the barges (up to six on the Rhine or 70 on the Mississippi) are lashed together and in turn lashed to the towing knees of the push-tug, the tow thus formed acts like a single vessel, ensuring more efficient use of energy and requiring much less labour.

Under way, the push-tow offers an awe-inspiring spectacle for the observer. There is undeniably something magical in the way the snub-nosed, squat push-tug powers, steers and controls its tow, easing up through tight bends on the Mississippi, momentarily backing up to let the current swing the whole tow round, before plying on upstream at a steady 10 km/h or more. Equally impressive is the sight of even a two-lighter tow sweeping down through the bends of the Rhine gorges at double that speed, its angle of drift so wide that the skippers of upstream-bound barges need all their nerves to hold their course, confident in the push-tug master's ability.

The advantages of push-towing as a provider of low-cost transport (pipelines can do better, but only for fluid cargoes) are now widely recognised, as is testified by the almost automatic adoption of this technique in plans for developing river navigation in third-world countries. And yet, push-towing has attracted very little attention as a subject in itself.

It must be admitted that there is nothing very striking in the basic principle of pushing, rather than pulling, tows, and this is especially true in the United States, the country with the longest history of regular push-towing operations on inland waterways. As will be seen, pushing came naturally on American rivers. More ingenious, or at least more innovative, were the various applications of the technique at different times in Britain, and its enormously successful introduction on the main rivers of Europe (essentially the Rhine and Danube), which because of their very different barging traditions and navigational conditions remained the almost exclusive province of conventional tows until the late 1950s.

In any event, it is hoped that this new 'international' look at push-towing will serve a useful purpose, perhaps stimulating further research, however uncertain the links between development of the technique in different countries.

Let us first of all briefly consider the early days of inland water transport on the great American rivers. For the pioneers settling in the 'wilderness' between the Allegheny mountains and the Mississippi river, the water courses were the only available means of transport during the late autumn and winter, when the trails and partially-cleared roads ('traces') became impassable. New Orleans, near the mouth of the Mississippi, became established from the early 19th century as the main market for their produce. The settlers had few tools and little nautical know-how, but in this vast region of largely primeval forest there was no shortage of timber. They put together rafts and the most elementary type of barge, flat-bottomed and box-like, covered from bow to stern, to carry their grain, bacon and whisky south. These vessels, called 'flat-boats', or 'boats', or 'broad-horns', were propelled downstream by the currents, their handlers simply striving by means of large sweeps to keep them in the channel. The boats were broken up at New Orleans, the lumber going into buildings or later being used as fuel.

As far east as Pittsburgh, where the Mississippi's main tributary, the Ohio, is formed by the Monongahela and Allegheny rivers, families migrating west bought or built flatboats for the one-way trip down river to what they hoped would be a prosperous new existence. In 1783 a boatyard at Elizabeth, Pennsylvania, 32 km. upstream from Pittsburgh on the Monongahela, was advertising for sale boats of 'every dimension'.

The need for a cargo vessel that could move upstream as well as down for transport of agricultural produce to market towns became apparent at an early stage. This need was met by the 'keelboat', a vessel between 12 and 25 m. long, with narrow beam (two to three m.) shaped bow and stern, and a keel (in the form of a long beam about 10 cm. square fixed to the bottom) which held the boat on course when under tow, and provided some protection in case of grounding on rocks, snags or sandbars. The strongly-built keelboat, carrying between 15 and 50 tonnes, was propelled upstream by between six and 18 men using long, iron-tipped poles, aided by square sails when the wind was right. This type of vessel was used notably for carrying grain to milling and market towns such as St Louis, on the Upper Mississippi. By the early 19th century, hundreds of keelboats were operating on the Mississippi and its tributaries (referred to simply as the 'western rivers'), in addition to the steady stream of flat-boats on their one-way trips.

In May 1811, just four years after Robert Fulton's pioneering steamboat trip on the Hudson river from New York to Albany, the river steamboat *New Orleans* was launched at Pittsburgh and set off downstream, arousing much excitement at all the riverside towns. However, this maiden voyage was hardly a great success. Technical incidents combined with the navigational hazards (notably shifting sand bars) meant frequent interruptions, and New Orleans was not reached until January 1812. Evidently, the deep hull design was well adapted to the Hudson river, but not at all suited to the shallow-water conditions on the upper rivers. The *New Orleans* did not return from her maiden voyage, but continued in service on the deeper waters of the Lower Mississippi until she was lost by sinking in 1814.[1]

In 1816 Henry Miller Shreve, an engineer–boatman–inventor from Shreveport, Louisiana, launched his steamboat *Washington*, built to an entirely different design which soon became standard on the western rivers. The straight-sided, flat-bottomed shallow draught hull was designed to provide buoyancy, rather than as space for cargo and engines. The engine was also very different from the low-pressure type (designed and built by Boulton & Watt in England) that had equipped the *New Orleans*. The new type was a high-pressure, non-condensing engine, using steam at about 100 lb. per square inch in a relatively small, horizontal cylinder. The 100-hp. engine weighed less than five tonnes, and drove the paddle wheel shaft directly.

With two paddle wheels, one on each side, set slightly astern of the boat's centre of gravity, there was an obvious advantage in having separate engines for each wheel, and all but the earliest boats were thus provided, ensuring the high degree of manoeuvrability that was essential for negotiating the twisting channels common at low river stages. By 1840 the shallow-draught side-wheel steamboat had fully evolved, but only rarely did it take flatboats or keelboats in tow. Safety must have been an important factor here, since towed vessels would regularly be running on to snags or sand bars.

The next stage in development of the steamboat brings us closer to the push-tow. This was the adoption of stern-wheel propulsion from about 1850, heralding the 'golden age' of steamboat operations on the western rivers. With a stern wheel the hull could be made wider and of shallower draught for a given total displacement without increasing the overall width of the boat. The technical difficulty which prevented its earlier introduction was that of providing adequate structural support for the heavy wheel at the stern. This having been solved, there remained the problem of making the stern-wheeler as manoeuvrable as the side-wheeler. Double wheels and split wheels were tried, but abandoned in favour of a system of multiple balanced rudders placed immediately forward of the paddle wheel and extending back under the wheel for some distance. These were particularly effective when backing, although additional 'monkey' rudders had to be fitted aft of the wheel to improve forward steering power.

With stern-wheelers, 'breasting-up' became possible, whereby barges built like flatboats or adapted from former steamboat hulls, or more rarely keelboats, were made fast alongside the steamboat, to be towed in a rigid formation. Steamboat operators were thus able to increase their carrying capacity to meet the fast-expanding transport demand. However, the number of barges which could be taken 'on the hip' was limited, one on each side being the most common configuration.

Already by the time the stern-wheeler was being introduced, flatboats loaded with coal for export were being formed into rigid tows at Pittsburg, so that even without mechanical propulsion the costs of transporting the 'black diamonds' down the Ohio and Mississippi could be reduced. These were in a sense 'captive' fleets just waiting for suitable motive power to be developed.

It was in 1845 that the first recorded movement was made in which a steamboat made fast to one of these fleets and pushed it to its destination at Cincinnati. The steamboat was adapted for this task simply by bolting timbers on to the bow, to butt against the barges. There were clearly some difficulties involved, for despite the obvious economic benefits to be obtained by steam propulsion of these coal fleets, it was not until 1851 that push-towing became accepted practice. The key to safe handling of these large, heavy tows was the so-called 'flanking' manoeuvre, already described earlier, whereby the tow is eased round bends. Downstream, the tow is allowed to drift with the current, the towboat operating in reverse to check forward motion and acting as a powerful rudder to position the stern of the tow as required relative to the direction of flow. The stern-wheel design of steamboat, with its high steering power when backing and its reserve power (obtained simply by increasing the steam pressure) made it ideal for push-towing, and the stern-wheel towboat became the workhorse allowing rapid growth of the towing industry during the 1850s. The arrangements for lashing the barges to the towboat were unsophisticated, and in hydro-dynamic terms these tows were not very efficient, but the history of push-towing had begun.

The Civil War disrupted river traffic between 1861 and 1865, and then railway competition began to make itself felt, but the long-distance downstream coal hauls continued until 1915, by which time new coalfields had been opened up farther downstream.

It was during the early days of American push-towing when William Bartholomew, engineer of the Aire & Calder Navigation in north-east England, designed an integrated transport and cargo-handling system involving push-towing of a very different nature. Trains of barges hauled by steam tugs had been successfully operating on the canal since 1831, and by 1855 two-thirds of the company's own carrying mileage was steam-hauled. However, in 1861, it was feared that the parallel railway between Hull and Leeds would be able to offer cheaper rates for coal haulage. This gave Bartholomew the opportunity to apply his invention, patented in February 1862. What he proposed was not so much a push-tow as a compound articulated vessel made up of separate compartment boats, with a shaped bow compartment in front and a steering and propulsion compartment at the rear. Each of the 25-tonne compartments was to be bodily lifted out of the water and unloaded in minutes by a specially-designed hydraulic hoist in Goole docks. The most original aspect of Bartholomew's invention was the steering system, involving wire ropes passing from fixed points on the bow compartment through suitable guides on each of the intermediate barges and round a capstan or windlass on the stern propulsion unit.

> By turning the capstan or windlass, the chain on one side of the compound vessel is wound up, and the chain on the other side is slacked out to a corresponding extent, and thus the compound vessel is drawn into a curved form, each barge pivoting about its projecting cutwater.[2]

The first seven-boat train of 'pans' or 'Tom Puddings', loading 168 tonnes, its overall dimensions corresponding to those of the recently-enlarged locks on the canal, started commercial working in 1865. It was followed by others, but as traffic increased it became apparent that longer trains could be operated if push-towing were abandoned. The more flexible system which has now been operating for about 100 years involves placing a tug at the head of the tow, with a false bow ahead of the first compartment boat, each boat being fitted with spring loaded buffers at one end. Under tow, a train thus curves round a bend and straightens automatically without the need for steering. Trains of 19 boats loading 700 tonnes have now been regularly run since conventional towing was adopted, with occasionally as many as thirty or even forty. The possibility of longer tows may not have been the only reason for the abandonment of push-towing. In 1890 W. H. Bailey, of Salford, suggested that the pushed barge train as patented by Bartholomew was

> certainly very likely to produce mishaps and damage both to banks and to the train itself.[3]

How well qualified Bailey was to make this remark is uncertain, but he is not the only observer to have criticised the push-towing/steering system. Staff at the canal basin serving St John's Colliery, Normanton, spoke of 'too much flexibility' in the steering system.[4]

Yet another form of push-tow was introduced on the Seine in France in the early days of steam navigation, thus in fact predating even the early American coal tows. From 1822, the carrying company Frossart & Margeridon applied its system involving

> two types of craft, one powered, pushing in front of it the other type, designed to carry merchandise ... the carrying vessel had its stern shaped to fit exactly into the bow of the powered vessel, with which it was linked by a combination of means.[5]

An early drawing shows one of these units, the *Etna*, under steam on the Seine. The main features are stern-quarter paddle wheels, driven by a steam engine situated well forward on the propulsion vessel. The unit appears to be steered by a large single rudder with a long tiller. The limited steering power thus obtained would have been a less serious constraint on the Seine, a naturally deep river with little current, than on the upper rivers of the Mississippi basin.

Another French innovation was the canal push-tow, tried on the Canal de la Marne au Rhin in France from 1855 to 1875. To avoid immobilising the propelling power of a barge during loading and unloading, an enterprising boatbuilder designed and built an integrated unit made up of a twin-screw 'tug' driven by steam engines and a single barge, the rounded bow of the tug fitting into the concave stern shape of the barge.[6] The idea was later taken up on the Seine for the close-quarter handling of lighters in the port of Rouen. To this effect, the *chaland rouennais* was designed with a concave stern, accommodating a steam push-tug of completely circular shape. The system was in use in the late 19th and early 20th centuries, and can be seen on old postcards. The units thus formed were in a sense forerunners of the North American integrated tug-barge combinations designed for deep-sea haulage. In both cases the finality is in effect a single vessel, with propulsion unit separated from carrying capacity, as opposed to a multiple-barge tow. The hull of the tug was shaped so that the screws, partly enclosed, would work in full flow conditions. Otherwise the propeller, of large diameter and necessarily set close to the water surface, would thrash a mixture of air and water, with a corresponding waste of energy. This must have been one of the first applications of the 'tunnel-stern' principle which was to become fundamental for efficient screw propulsion on shallow inland waters.

The main problem on the Canal de la Marne au Rhin was the small size of the locks, 38.50 by 5.20 m. Provision for the tug-barge combination to fit inside the lock chamber meant a substantial loss of cargo capacity by comparison with dumb barges hauled from the towpath. Otherwise the barge and tug would have to be split to negotiate each lock separately, making progress prohibitively slow, especially on a canal with an average of one lock every two kilometres. Bank traction was thus shown to be more economical on the French canals, and electric locomotive traction survived until 1970. Even so, the idea of push-towing barges as big as the locks would allow has continued to be advocated. In 1950 E. C. Jones, whose original push-towing operation in England will be examined later, suggested the use of separate push-tugs on each pound, shuttling dumb barges from one lock to the next.[7] Today, the same approach is proposed for the canalised river Douro, in Portugal, due to be opened to navigation by the time this book is published. Overall capacity of the waterway could be increased from seven to 10 million tonnes per year by the use of barges filling the locks completely, handled by separate push-tugs on each of the five pools.[8] Interestingly, the analytical model study showed that this increase in traffic could be achieved with virtually the same number of push-tugs (about twenty-five).

The British and French applications of the push-towing technique we have just seen both ran against the general trends in carrying by barge in these countries. If Bartholomew's system was the more successful, this must be attributed to the relatively large dimensions of the locks on the Aire & Calder Navigation, and to the fact that it included provision for economical loading and unloading of the Tom Puddings. In any event, it was a 'one-off' design with no subsequent application elsewhere on the British waterway system.

Meanwhile, push-towing was becoming more widespread on the western rivers of the United States. In particular, log rafts began to be pushed after the Civil War, and Samuel Clemens (alias Mark Twain), returning to the Mississippi in the 1880s, noted significant changes since his steamboat piloting days from 1856 to 1861. On the upper river

> we met massed acres of lumber-rafts coming down, but not floating leisurely along, in the old-fashioned way, manned with joyous and reckless crews of fiddling, song-singing, whisky-drinking, breakdown-dancing rapscallions; no, the whole thing was shoved swiftly along by a powerful stern-wheeler, modern fashion; and the small crews were quiet, orderly men, of a sedate business aspect, with not a suggestion of romance about them anywhere.[9]

Stern-wheel steamboats with improved hull designs and multiple rudders had been found adequate for handling the coal-boat tows from Pittsburgh, but lumber rafts were a different matter, being both larger and less rigid than the tightly-lashed tows of coal barges. During the 1860s rafts were occasionally towed by steamboats in the conventional manner, but the operators soon started building steamboats specially adapted for push-towing. Provision was made initially for curving the rafts through bends, in much the same way as Bartholomew's barge-train on the Aire & Calder. Lines attached to each side of the raft were run through a capstan on the bow of the towboat, and by shortening the line on either side at will, the raft could be swung round. However, this could only work effectively with a limited size of raft. Instead, it became common to use a 'bowboat', a small boat attached crosswise at the bow of the raft, so that by backing or advancing it could assist the towboat in guiding the raft. With this arrangement, it was possible to push much larger rafts, up to 1,500 feet (460 m.) long and more than 100 feet (30 m.) wide, although at certain difficult or narrow locations, such as rapids or railway bridges, rafts continued to be 'double-tripped', that is, split into halves lengthwise, each half being taken through separately, to be joined again beyond the difficult stretch. The rafting business reached its peak around 1890.[10]

The turn of the century was a period of decline in American waterborne transport, essentially because of unregulated railway competition and the always limited navigability of the western rivers. It must be emphasised that all the developments described so far took place on shallow-draught navigations, however substantially these may have been improved over the years (notably under the 1824 Act 'to Improve the Navigation of the Mississippi and Ohio Rivers'). Only the Monongahela river had been canalised with fixed weirs and locks, the works being completed in 1890, providing a navigable depth of 6 feet (1.80 m.). This, together with completion of the first weir and lock on the Ohio downstream of Pittsburgh, encouraged heavy movements of coal by barge from the mines along the Monongahela to the steel mills in the Pittsburgh area. This push-towing operation continued to grow during the period of overall decline, and provided the impetus for further improvement of push-towing techniques.[11]

Steamboat efficiency was increased by the introduction of feed-water heaters and multi-flue boilers, while condensers improved the performance of the compound steam engine. Steel-hulled towboats and steel barges of more efficient form were introduced from about 1910. The diesel engine was first used for towboat propulsion about 1916, offering prospects of a substantial increase in efficiency, and experiments were made with screw propulsion.

At the beginning of the modern era, after World War I, push-towing on the western rivers of the United States had thus come a long way technologically from its rather rudimentary beginnings in the 1850s, and the water transport industry was able to respond rapidly to the enormous increase in transport demand which then occurred. This sudden change in fortune was due to the great difficulties faced by the competing railroads in the post-war period. Both the Federal Government and private industry realised the dangers of over-dependence on a single transport mode, and deliberately set out to promote wider use of water transport. This effort was not limited to the continuing process of improvement of the rivers, particularly their depth. In 1924, the Federal Government also established the Inland Waterways Corporation, which later became the Federal Barge Line. This corporation built a fleet of steam-powered towboats of both stern-wheel and propeller types, and inaugurated a common carrier service on the Mississippi.

The 1920s were a period of growth in traffic and change in techniques. Diesel propulsion was gradually taking over from steam, a process delayed only by latterday improvements in steam propulsion technology. More significantly, the substantial river improvement and canalisation works put in hand by the Federal Government, notably the complete canalisation of the Ohio river with 43 locks and weirs, finished in 1929, were gradually establishing a network of channels offering a guaranteed depth of nine feet (2.75 m.). This was to toll the knell of stern-wheel propulsion, essentially adapted to shallow-draught conditions, although here too the change took place only gradually, because depths remained inadequate on many rivers and because the early tunnel-stern towboats were deficient in backing and flanking power compared with stern-wheel boats of similar horsepower. The last stern-wheel steamboat (except for a few built under exceptional wartime conditions) was built in 1940, long after the beginning of modern large-scale push-towing operations.

In Europe, it was the Danube shipping company Bayerischer Lloyd which first tried to adopt the push-towing technique. In 1928 the company adapted its twin-screw 350-hp. steam tug *Isar* to push two laden 700-tonne barges, trials being conducted initially on the still water pool formed by Kachlet lock upstream of Passau (incidentally, the first lock to be built on the Danube), and then on the free-flowing river downstream through Austria. The results were sufficiently convincing for the company to place an order for a purpose-built push-tug. The *Uhu*, 42 m. long, powered by two vertical shaft Voith-Schneider propellers developing 600 hp., was launched in 1930, and started operating commercially on 10 October of that year. Tests had shown manoeuvrability and towing efficiency to be excellent, with prospects of transport costs being reduced by a third, but they had also revealed mechanical and structural design faults, which were never completely eliminated. Water seeped through the propeller shaft seals into the sump, and precision of the hydraulically-operated rudder was affected by oil overheating. Also, under certain conditions, water came into the vessel through the propeller wells. Remedial action was taken, but in seven years the push-tug never gave more than six months of uninterrupted service. The experiment was given up in 1939, and the *Uhu* converted to serve as a conventional twin-screw tug (until she was sunk by a mine in the Iron Gates section of the Danube in 1944).[12]

As push-towing has since been adopted on the Danube on a large scale, it is possible that the failure of this early operation was due to inadequacy of design of the push-tug *Uhu* itself rather than to any inherent failing of the push-towing principle applied to this river. In particular, it should be noted that the Voith-Schneider propulsion system has rarely been applied since on 'line' push-tugs (as opposed to the harbour vessels designed for moving barges around at their loading and unloading terminals). Even so, it is clear that on both Rhine and Danube conventional towing techniques were becoming more refined and efficient, and skippers and pilots showed great skill in negotiating the relatively tight bends of these often fast-flowing rivers. Upstream and downstream tows regularly had to cross each other, using the 'blue flag' procedure, the upstream tow keeping to the inside of the bends to obtain the full benefit of of reduced current speed. In view of the delicate manoeuvres thus performed by conventional tows on rivers with dense traffic (especially the Rhine), there was an understandable reluctance to accept rigid push-tows and the further complications this would inevitably entail. As late as 1957, a Dutch engineer remarked that large-scale application of push-towing on the Rhine was a 'very unlikely' prospect, and that interchangeability of the entire Rhine fleet would 'no doubt be seriously disturbed by a partial switch-over to push towing'.[13] (See figure 3.1 opposite.)

In the post-war period, push-towing was an accepted practice only on the lower Danube below Komarom, in Hungary. Soviet, Yugoslavian or Bulgarian barges were occasionally formed into push-tows of an unusual 'arrow' configuration, whereby rigidity is obtained partly by breasting-up. With this formation, even a conventional tug without towing knees could be used to push six barges.[14]

While conventional tows reigned on the Rhine and Danube, interesting developments were taking place on the Belgian Congo (now Zaïre). Breasting-up had been regular practice from 1914, a self-propelled barge pushing two dumb barges 'on the hip',

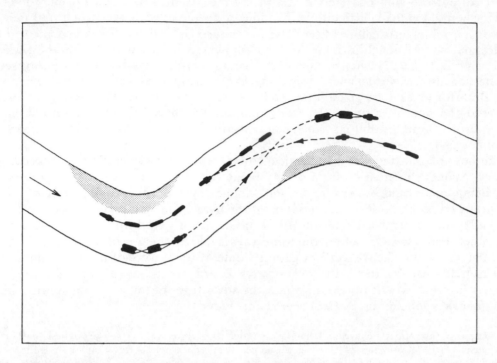

Figure 3.1. Conventional tows cròssing the fairway on the Rhine, as required by the current. The view that rigid formations could not mix safely with conventional traffic long delayed the development of push-towing on the Rhine.

as thé Mississippi packet boats had done. Later, after a period in which conventional line towing was tried on both the Congo and its main tributary, the Kasaï, push-towing was introduced in 1935, using existing tugs and barges, the formation being similar to that just mentioned on the lower Danube. The rapid increase in traffic after World War II meant that new equipment had to be purchased, and the navigation authority took this opportunity to invest in modern push-tugs and adapted barges. By 1950, push-tows of 3,000 to 4,000 tonnes were plying the Congo and the Kasaï, and in 1953 totally integrated tows were also purchased.[15] In this type of tow, already introduced in the United States, the flexibility offered by the use of barges of symmetrical shape (identical bow and stern forms) is sacrificed in favour of fixed-formation tows, in which only the extremities of the overall tow are shaped. Thus the outer barges are shaped at one end and straight at the other, while intermediate barges are straight at both ends. Besides increasing cargo capacity, this fixed formation offers less resistance and hence reduced energy consumption and greater speed for a given horsepower rating. Integrated tows are ideally suited to bulk flows of a single commodity between fixed terminals, and under these conditions towing efficiency can be improved by 30 per cent.

Curiously, it is to Britain's small waterways that we must now turn to see the first significant post-war development in Europe. In fact, the first attempt to navigate a vessel by push-towing occurred in 1939, on the Thames between Boveney lock and Bray, when a tug had difficulty in handling a deep-loaded barge which had failed to answer its rudder in shallow water. The tug moved behind the barge and pushed it to its destination. Although pushing was thus shown to be effective, the method was not used again until 1947, when an adapted Thames towing tug started continuous service on the canalised river Thames, improving barge turn-round with reduced manpower. The Brentford-based barge-building and towing company E. C. Jones & Son Ltd. patented the new idea under the name 'Bantam' both in Britain and abroad, and its first purpose-built push-tug, small and relatively inexpensive, entered service in October 1948.

The bow of the tug with two vertical towing knees set close together was designed to push against the stem or stern of any barge (even Thames 'swim-end' lighters), the unit being made rigid by wire ropes and winches over the forward quarter of the tug and attached to bollards on the quarters of the barge. The company subsequently built 90 similar push-tugs, used typically to push barges on gravel pits, or dredging hoppers and other maintenance barges on waterways such as the Leeds & Liverpool Canal, the tidal Trent and the Weaver Navigation. Some of them were of narrow beam (7 feet, 2.10 m.), for use by the British Waterways Board on its narrow canals. These were all single-barge tows, demonstrating that the advantages of push-towing are not limited to large-scale applications. A 1961 report concluded:

> These pusher units behave and handle in much the same way as self-propelled craft and because of the good lines of the tug the control is very much better than with traditional towing, and often equal to or better than that of a self-propelled craft. Consequently, the introduction of push towing has not brought with it any problems of manoeuvrability in pushing in restricted waterways.[16]

A curious aspect of this push-towing operation on the Thames was that it contravened a Thames Conservancy bye-law. Not wanting to condemn what 'might possibly prove to be a new and improved method of propelling barges on the river', the Conservators willingly turned a blind eye to this infringement, until one of their inspectors caught the *Bantam I* on 21 November 1949 simultaneously pushing one barge and pulling another! The Conservators felt this was carrying things too far, and claimed that under these conditions the tug was 'not readily navigable or sufficiently safe'. In a test case the company and its unfortunate skipper (who declared he was acting under instructions from his employer and his union) were prosecuted and nominal fines imposed.[17] This quirk of justice did not prevent further development of the Bantam system, although the overall decline in waterborne transport, related to the lack of investment in waterway improvements, certainly retarded the development of push-towing for regular bulk traffics.

American-style push-towing was introduced into France in 1955 by the Lambert cement works, with their push-tug *Poussah* and two 1,200-tonne lighters carrying limestone on the Seine downstream of Paris. Encouraged by the success of this first tow, the Sablières de la Seine, a large sand extracting undertaking, decided in early 1957 to adopt push-towing for transport of its materials over the 180 km. from Les Andelys to Paris. There were then seven locks on the lower Seine, the negotiation of which by conventional pull-tows was extremely laborious, thus highlighting a further advantage of push-towing on canalised rivers. The company rapidly built up a fleet of four push-tugs and 36 lighters, capable of carrying 400,000 tonnes a year.[18]

It was also in 1957 that push-towing was tentatively introduced on the Rhine, two separate initiatives bearing fruit almost simultaneously. These involved respectively the large French undertaking Communauté Française de Navigation Rhénane (CFNR) and a German/Dutch venture. Different accounts refer to one or the other as having been the 'first' to apply the technique on the Rhine. What is more important is that by this time the advantages of push-towing as demonstrated by American experience over several decades had become more convincing than the reactionary arguments of the die-hard pull-towing lobby on the Rhine,

The CFNR, concerned to acquire experience of push-towing by means of full-scale tests before converting its entire towing fleet, adapted one of its tugs, an 850-hp. vessel equipped with two Voith-Schneider propellers, and started converting 1,600-tonne barges into push-tow lighters by stripping out their wheelhouses, living quarters and rudders. The push-tug was commissioned in early September 1957, and its first two-lighter tow entered the port of Strasbourg on 18 September. A second Voith-Schneider tug was then adapted, followed by a 2,400-hp. triple-screw tug. The latter was intended specifically to determine the maximum push-tow dimensions feasible on the upper Rhine. This push-tug with two lighters, forming a tow 242 m. in length, successfully navigated upstream through the difficult section of the Rhine from Lauterbourg to Strasbourg. On the other hand, heading downstream with four lighters (two abreast ahead of the push-tug and one on either side of the latter, making overall dimensions of 170 by 30.50 m.), difficulties appeared in the two bends at Drusenheim and Greffern. Entering the bends, the pilot adopted a fairly wide drift angle, which is normal practice, but then had to race through the bends at breakneck speed to keep steerage way relative to the fast current and avoid grounding on the outside bank. Flanking, Mississippi-style, was then attempted to keep better control of the tow, but here too the full power of the converted tug was required to extricate the tow from what CFNR's engineers admitted became a 'rather tricky situation'. The four-lighter push-tow was manifestly too dangerous, and CFNR decided to operate only two-lighter tows until the substantial improvements to Rhine navigation, completed in the 1970s. allowed operation of four-lighter tows throughout the length of the river, including the gorges section, after the notorious Binger Loch had been blown up in 1974.[19]

By contrast, navigable conditions downstream of Duisburg were always far more favourable, and the German company Raab–Karcher Reederei (Duisburg) took the risk of ordering its first push-tow from a Mainz yard without conducting preliminary full-scale tests. The push-tug *Wasserbüffel* (water buffalo), 36 m. long, powered by two 630-hp. Deutz diesel engines, was launched in October 1957, a few weeks after CFNR's first tow started working. The push-tug, with four 1,300-tonne lighters, was assigned to the Rotterdam–Duisburg iron ore traffic, operated in association with the Dutch firm NRV Vulcaan. The *Wasserbüffel* and sister vessel *Nashorn* (rhinoceros) were the first in a long line of push-tugs successively assigned to this ever-important traffic.

Since these beginnings, push-towing has grown at an astonishing pace, not only on

the Rhine but also, since the late 1960s, on the Danube, the Rhône, the Dunkirk–
Valenciennes waterway, the Elbe and all the main rivers in Eastern Europe, the Po in
Italy, and of course on the main rivers of South America, Africa and Asia. In all cases
it is basically the American principle of the rigid push-tow that has been adopted,
although William Bartholomew would have been pleased to learn that, even in these
modern times of easy communication and worldwide standardisation of technology,
there have been attempts to operate flexible, or 'deformable rigid', push-tows similar
to the barge-train he patented in 1862.

Thus British Petroleum's associate company, Société Française des Pétroles BP,
commissioned on the Seine in 1964 an articulated tow made up of a 1,700-hp. push-
tug and two tanker barges with a capacity of 4,000 tonnes, the total length being
about 150 metres. The forward barge was coupled by two steel cables to hydraulic
rams on the bows of the push-tug, the intermediate barge being sandwiched in
between. Rubber retaining bolsters set in vertical recesses on each unit prevented any
direct contact between the hulls. The tow was articulated, that is made to curve
as required, by applying hydraulic pressure in the coupling rams and operating the
rudders.[20]

Following a similar principle, the West German company Cassens (Emden) built a
self-propelled articulated barge made up of four units with a total length of 162
metres.[21] This 3,000-tonne tow, powered by two 240-hp. diesel engines, one on each
of the end units, was designed to take advantage of the full length of the locks on the
West German canals, while negotiating without difficulty the relatively sharp bends on
these canals (originally built for 1,000-tonne barges no more than 67 metres in length).

Flexible tows were also tried in the Soviet Union, on the Volga, the Irtysh and the
Amur, and in Austria comprehensive model tests were undertaken to determine the
feasibility of introducing such tows on the upper Danube between Vienna and the
West German port of Regensburg. This section had 20 bends of 500 m. radius or
less, in which the path swept by a six-lighter tow, for example, would be 80 metres
wide, or 30 metres more than the low-water channel width! It was thought that the
only way push-towing could be adopted would be if the tows were made to swing,
with a hydraulic ram arrangement like that on the French and German tows just
mentioned.[22] The tests produced favourable results, but the authorities clearly felt
there were too many risks involved, and continued to build traditional pulling tugs
and barges.

None of these flexible push-tows worked for more than a few years. The failure of
all these experiments must be attributed to the inherent weakness of the links, which
were bound to be a source of trouble. On the main rivers, police regulations now
specifically prohibit the operation of deformable rigid push-tows.

There has, however, been in Europe one lasting departure from normal American
practice. This is the operation of rigid units made up of a self-propelled barge and a
dumb barge pushed ahead of it. This technique is used typically by small operators
who are able to purchase dumb barges relatively cheaply and can thus accept larger
consignments when the market is favourable. Mini-push-tows made up of a self-
propelled 38.50-m. *péniche* and a similar unpowered craft, together loading between
500 and 700 tonnes, have become a common sight in the Seine basin and in northern France. They were a natural product of the Canal du Nord, completed in

1964, which has locks 92 m. long but only 6 m. wide. The master plan for the French waterways includes upgrading works specifically designed to open up new routes for the mini-push-tow. On the Rhine, too, some operators have invested in modern, high-capacity self-propelled barges designed to push one barge ahead. This is essentially to give the maximum possible flexibility in what has become a very difficult market.

Another factor acting against bulk push-towing in Europe is the tenacity of the private barge owner-operators, who achieve excellent results by working intensively (up to 18 hours per day), carrying out maintenance on board and using the whole family. These efficient motor barges make it difficult for the big company push-tows to compete on smaller waterways such as the Moselle or the Main, where push-tows are limited to two lighters, and all the more so on the Neckar, where only one lighter can be pushed. Some barge-skippers' unions have nevertheless always seen industrial push-towing as a threat to their livelihood, and have occasionally resorted to militant action in attempts to prevent certain traffics from being handled by push-towing fleets outside the normal freight exchange system.

Push-towing is also deprived of certain traffics by railway rate-cutting. Thus one of the main projected traffics through the new Elbe Lateral Canal, iron ore carried in push-tows from Hamburg to the Salzgitter steelworks (served by a branch of the Mittelland Canal), was lost at the last minute when the German Federal Railway proposed a rate which shipping circles claim does not even meet the marginal railway operating costs.

Even so, there remain numerous bulk traffics on European waterways which are most economically handled by push-tugs and lighters. The Rhine-based push-towing fleet, made up of about 180 push-tugs and 1,000 lighters, accounts for a third of the tonne-kilometres carried on the river, and this proportion is increasing as waterway improvements gradually open up new routes for push-towing. Thus the Scheldt–Rhine Canal, opened in 1975, gives six-lighter push-tows access from the Rhine to the Belgian port of Antwerp. Large-scale upgrading works in progress on the West German canals and on the waterway links between Belgium and France are also specifically designed to extend the operating possibilities of high-capacity push-tows.

Although slow to adopt push-towing after the failed experiment with the *Uhu* in the 1930s, the Danube shipping fleets quickly made up for lost ground during the 1970s. The Yugoslavian River Shipping undertaking was the first to bring into use modern triple-screw push-tugs, and since 1972 has regularly run 12-lighter tows loading more than 16,000 tonnes between Reni (USSR), km. 130, and Kladovo (Yugoslavia), km. 930.[23] Tows have to be split up to pass through the Iron Gates section above Kladovo, but despite this obstacle (and the incomplete canalisation of the upper reaches through Austria to Regensburg), push-towing now accounts for half of the total fleet capacity on the river, with about 180 push-tugs and 1,200 lighters.

Back in Britain, push-towing was limited to the handling of single barges, as already described, until in 1964 the long-established canal carrier Hargreaves formed a subsidiary company Cawoods Hargreaves Ltd., to operate a push-towing fleet made up of nine push-tugs and 35 modern compartment boats, each carrying more than 150 tonnes of coal. Hargreaves had won a contract for carriage of more than a million tonnes of

coal per year to the new Ferrybridge 'C' power station on the Aire & Calder Navigation, and this traffic continues today.

Less fortunate was the push-towing operation introduced by the British Waterways
Board on the waterways of north-east England in conjunction with BACAT barge-
carrying system. Technically, the system was excellent, and offered British industry the

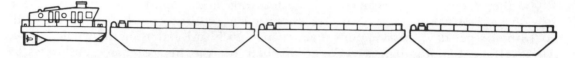

Figure 3.2. A 3 x 140-tonne push-tow built for operation on the commercial
waterways of north-east England.

prospect of cheap barge transport without break of load between inland terminals on
the Continent and in Britain. While he was on the Board in the early 1960s, Charles
Hadfield, to whom this book is dedicated, played a key role in promoting this concept.
The barges, specifically designed for the relatively small dimensions of Britain's commercial waterways, crossed the North Sea in the mother ship *BACAT I*. Unfortunately,
Hull dockers were not prepared to see the BACAT traffic slipping past their noses, and
blacked the operation until it had to be abandoned in 1975. The BWB continue to
operate the fleet of push-tugs and barges on domestic traffics, although obviously with
more limited potential than if the through hauls to the Continent had been maintained.
In passing, it should be noted that optimum use of the holds of barge-carrying vessels
necessarily implies box-shaped barges, which made push-towing the obvious choice of
propulsion mode for the inland legs of the hauls.

Despite this setback, push-towing of 500-tonne consignments on the waterways
of north-east England remains highly competitive for bulk traffics, as was again
demonstrated in 1980 when Inland Waterway Carriers Ltd. (formed by Cawoods
Hargreaves and the BWB) won a contract to transport 670,000 tonnes of minestone
waste per year on the Aire & Calder Navigation and the Humber. The compartment
boats for this traffic are to be fitted with self-discharging apparatus driven by mains
power at the unloading terminal.

The extent of push-towing operations on Britain's relatively small commercial waterways (Table, p. 60) shows that the usefulness of push-towing is not proportional to the
capacity of the tow. Conditions in the Mississippi basin, where water, rail and to a limited
extent road are competing for hauls of thousands of kilometres, are totally different from
those in the Rhine basin or in north-east England. In other words, the 3 x 150-tonne
barge push-tow in England, by offering a cheaper freight rate than British Rail or road
haulage firms, performs a function in the economy just as important as that performed
by the 40 x 1,000-tonne barge push-tow on the Mississippi. Until the recent upgrading
of the South Yorkshire Navigation, successive governments in Britain had made the
tragic mistake of writing off Britain's waterways as of minor significance, perhaps
misled by apparently unfavourable comparisons with other countries like the one we
have just made. Fortunately, this simple lesson in the economics of physical distribution
has now sunk in, and the importance of inland shipping is now widely recognised.

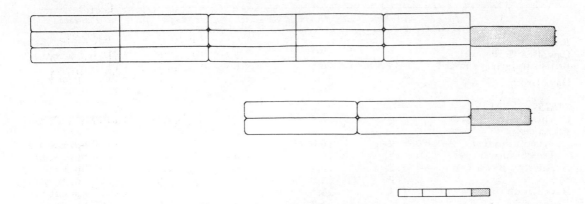

Figure 3.3. Comparison of standard push-tow formations on the Ohio (top), the Rhine (middle) and the commercial waterways of north-east England (bottom), as determined by maximum lock dimensions. The first two are formed with broadly similar units, the standard European barge being slightly larger than its American counterpart. However, the wide American rivers allow a much greater number of barges to be formed into a single tow. Note that the Ohio tow is here assumed to be formed with 'integrated' barges, which are bluff at one end and assembled in pairs end-to-end, thus reducing resistance. The table raises the interesting question of what standards should be adopted for Britain. The larger lock dimensions are those of the newly-upgraded South Yorkshire Canal, with dimensions of the Aire & Calder locks given in parentheses. The standard push-tow operating here today has to be split to negotiate the latter, while making far less than optimum use of the 700-tonne lock capacity on the former.

TABLE TO FIGURE 3.3

		Ohio	Rhine	N.-E. England
Lock dimensions	Length	365.80	195.00	77.00 (54.85)
	Width	33.50	24.00	7.00 (5.65)
Standard barge dimensions	Length	59.40	76.50	16.75
	Width	10.70	11.40	4.70
Approximate unit barge capacity: (depending on draught)		1,500 t.	2,500 t.	140 t.
Typical push-tug dimensions	Length	56.40	40.00	12.25
	Width	13.70	11.40	4.70
Tow formation — in locks		5 x 3	2 x 2	3 x 1
— in open river		5 x 6	2 x 3 (3 x 2)	
Tow dimensions in lock	Length	353.40	193.00	62.50
	Width	32.10	22.80	4.70
Approximate maximum tow capacity:	in lock	20,000 t.	10,000 t.	500 t.
	in river	40,000 t.	15,000 t.	

	Number	Length (m.)	Breadth (m.)	Draught (m.)	Remarks
BARGES					
Cawoods Hargreaves	35	17.1	5.3	2.5	170-t. capacity
British Waterways Board	32	16.8	4.6	2.7	140-t. capacity
Hays Group	(6	40.3	9.1	3.2	800-t. capacity
	(2	30.7	7.8	2.7	400-t. capacity
TUGS					
Cawoods Hargreaves	9	8.8	4.3	1.2	1 x 150-hp. engine
BWB classes					
Freight Pioneer	2	10.0	4.6	1.7	2 x 150-hp. engines
Freight Mover	2	15.2	4.6	2.7	1 x 420-hp. engine
Allerton Bywater	2	15.2	4.0	1.8	1 x 135-hp. engine
	(2	24.3	8.5	2.6	2 x 365-hp. engines
Hays Group	(1	25.5	6.7	3.5	1 x 650-hp. engine
	(1	21.3	5.3	2.6	1 x 365-hp. engine
E. V. Waddington	1	15.2	4.5	1.4	1 x 130-hp. engine
John Dean	1	n.a.	5.3	2.3	n.a.

British commercial push-craft, excluding dredging and gravel-pit craft.[24]

Push-towing may still be a long way from becoming accepted practice for deep-sea barge movements, with all the problems involved in accommodating wave-induced forces, but on inland waterways throughout the world, from Britain to China, from Siberia to Paraguay, it has established itself in a few decades as the most technically sound and efficient carrying system.

NOTES AND REFERENCES

1. Much of the detail of developments in the USA is taken from PERSON, J. L., & MYTINGER, R. E. Evolution of propulsion and control of cargo vessels on the inland waterways of the United States. *19th PIANC Congress, London 1957*. Section I, Q2. 59–81. The author is also most grateful to James V. Swift, Business Manager of *The Waterways Journal* and specialist in the history of American push-towing, for his own interesting account.
2. 1862 patent description quoted in HADFIELD, C. *Canals of Yorkshire and North East England*. Vol. 2. 1973. 368.
3. BAILEY, W. H. Notes on canal propulsion. Paper to *International Canal Congress, Manchester 1890*. Quoted in *Engineering*. 50. 15 August 1890.
4. MARSDEN, E. G. Letter: Push-towing and the ACN. *Modern Transport*. 98. October 1967. 11.
5. BEAUDOUIN, F. *La batellerie et Conflans Sainte Honorine*. Assn des Amis du Musée d'Intérêt National de la Batellerie, 1976. 74.
6. GRAFF, R. Paper on Evolution of types of propulsion . . . *19th PIANC Congress, London 1957*. Section I, Q2. 83–136.
7. JONES, E. G. Development of inland waterways: suggested future policy and towage methods. *Dock & Harbour Authy*. 31. December 1950. 261–262.
8. OVIDIO, J. N. C. L., *et al*. Feasibility and optimisation of a waterway on the River Douro. *24th PIANC Congress, Leningrad 1977*. Section I, Subject 1. 163–176.
9. TWAIN, Mark, *pseud*. [Samuel L. Clemens]. *Life on the Mississippi*. 1883.
10. HARTSOUGH, M. L. *From canoe to steel barge on the Upper Mississippi*. 1934. 120.
11. PERSON & MYTINGER. *op. cit*. 68.
12. PISECKY, F. Die grössten Schubverbände Europas auf der Donau. *Schiffahrt und Strom*. 1972. Issue 26/27.
13. GOEDKOOP, L. J. W., *et al*. Paper on Evolution of types of propulsion . . . *19th PIANC Congress, London 1957*. Section I, Q2. 177–217.
14. PRAGER, H. G. *Was weisst du vom Donaustrom?* 1962. 46–47.
15. LEDERER, A. Evolution du mode de propulsion . . . et du remorquage au Congo Belge. *19th PIANC Congress, London 1957*. Section I, Q2. 41–58.
16. BLENKHARN, A. Paper on Propulsion of vessels by push towing. *20th PIANC Congress, Baltimore 1961*. Section I, Subject 4. 177–188. *See also* BEAUMONT, Norman. Push-towing on inland waterways. *Modern Transport*. 98. July 1967. 46–48.
17. New barge propulsion method: test case by Thames Conservancy. *Dock & Harbour Authy*. 30. February 1950. 315.
18. VADOT, R., & DELMAS, G. Paper on Propulsion of vessels by push towing. *20th PIANC Congress, Baltimore 1961*. Section I, Subject 4. 115–176.
19. *Ibid.*, 125–128.
20. Pusher barge train for the River Seine. *Dock & Harbour Authy*. 45. February 1965. 305.
21. AGATZ, A. Trends in cargo handling and transport methods. *Dock & Harbour Authy*. 42. February 1962. 325–330.
22. VOLKER, H. The problems of pushed cargo-vessels on the Upper Danube. *21st PIANC Congress, Stockholm 1965*. Section I, Subject 3. 33–45.
23. PISECKY, F. *op. cit.*
24. INLAND WATERWAYS ASSOCIATION (pub.). *British Freight Waterways – Today & Tomorrow*. 1980. 21. (Updated by the author.)

CHAPTER 4

BRITISH PLEASURE BOATING ON THE CONTINENT (1851-1939)

PLEASURE BOATING became popular in Britain during the latter half of the 19th century. The 1860s had found the Victorian middle classes enthusiastic about travel and adventure. Thomas Cook had led his first tour to Switzerland. Boys were reading *Coral Island* and *The Gorilla Hunters*. Nelson was publishing — at the steep price of 12 shillings — accounts of *A Walk from London to John O'Groats*; Captain Hall was narrating the story of his *Life with the Esquimaux*; Mrs. Gerise Rogers had spent *A Winter in Algeria* and the Rev. Louis Noble had returned from *A Summer's Voyage to Labrador*. W. H. G. Kingston had written *The Cruise of the Frolic* for young yacht-loving people, and the *Boys' Own Book of Boats*. Lord Aberdeen had taken up canoeing and the Prince of Wales had become president of the newly-founded Canoe Club. Dozens of young men, spurred on by a sense of adventure, had had canoes built for them by Searle of Lambeth, and were actively planning and plotting voyages of exploration. Now there appeared a variety of accounts of voyages by gig, skiff, canoe, yawl and una boat, not only through England, but across Europe and the Middle East.

While the upper portions of the Thames with a light boat and fair weather were fine for many, the enjoyment for some Victorians was better fulfilled by roaming through distant countries with an element of danger and difficulty to be overcome. However, the actual source of enjoyment took different forms. Robert Mansfield, one of the first to write of pleasure boating in Europe,* remarked that human nature delighted in doing, seeing and being what fellows had not done, seen or been before and urged his readers to enjoy the pleasure of investigating 'the watery nooks and corners of old Europe'. 'Above all', noted Mansfield, 'it is the feeling of perfect independence and freedom in those extensive solitudes, where a human being scarcely ever sets his foot and where the silence was only broken by the dull roaring of a rapid, the booming of the bittern or the rattle and rush of the wings of the wild geese' which made their expeditions so enjoyable. John MacGregor, founder of the Canoe Club, described canoeing as a new mode of travelling 'by which few people and things are met with, while healthy exercise is enjoyed and an interest ever varied with excitement keeps fully alert the energies of the mind'.

Warington Baden-Powell preferred 'a stiff breeze, wild scenery, freedom of dress and action, and the perplexities of a strange language, to being guided by a tourist's handbook from a first-class Swiss railway carriage to the comfortable monster hotel with its obsequious English-speaking waiters'.

---

* A bibliography of European cruising, 1833-1939, follows this chapter.

This feeling continued to be echoed through the years. In 1878 James Powell writes of the joys of the Weser and its lovely scenery 'without hearing the dreaded names of Cook and Gaze'. James Molloy pointed out to those who might think small rivers and canals monotonous compared with the Seine or Loire that it was the very tranquility of the little streams and their unknown villages which made a pleasing change from the rush of the great rivers and their big cities.

Such was the philosophy of the Victorians, whose urge to promote rowing was engendered by the belief that this form of exercise is 'good for body and soul, and promotes sound and practical philosophy, in improving health and bodily vigour, and in sweetening the blood and tempers of men'.

Robert Mansfield

It was Robert Mansfield's *The Log of the Water Lily*, published in 1852, which was the first literary account to advise readers on how they might see 'the finest scenery in Europe in an economical, novel and delightful manner'. The *Water Lily* was a 36-ft. four-oared Thames gig, built by Noulton & Wyld of Lambeth, which became the first English rowing boat to be taken on an excursion on the continent. She was launched at Mannheim on 31 July 1851. Great was the astonishment of the Mannheimers at seeing five Englishmen clad in grey flannel trousers, white shirts and white felt hats with broad brims, 'insert themselves into an elongated walnut-shell'.

The crew rowed down the Rhine to Cologne diverting here and there to explore some of its tributaries without serious mishap. Indeed, so enjoyable was the trip that the following year a second *Water Lily* was shipped to Wurzburg and rowed up the Main and via King Ludwig's canal to Budapest.

Mansfield gives an interesting account of their leisurely progress through this 107-mile long waterway which, by joining the Main to the Danube, linked the North Sea to the Black Sea. 'Pulling along the canal was delightfully lazy work after our hundred mile tug up the Main'. From the canal's embankment the scenery was very pretty and alongside ran the river Regnitz, the high road, railway and telegraph. Mansfield noted that the canal's lack of commercial success appeared to be due to the shallows and rapid stream of the Main Navigation, the lack of water along the canal's summit level and the similarity of the productions at either end. Consequently they were told it barely paid the working expenses and they saw but few barges on it; however it had a 'most efficient and numerous staff of officers'.

At Dietfurt the lock-keeper, who had been one of the Bavarian volunteers under King Otho in Greece, told them the town was 'very bad, inhabited only by old peasants'. Mansfield noted the curious costumes worn by the ladies, 'gowns very short at the waist, sleeves wadded out to an enormous size'; admired the Altmuhl valley scenery with its castles perched on the tops of lofty cliffs, and delighted in the clarity and rapidity of the river in which they swam.

At Donaustauf the Princess of Thurn and Taxis invited them to leave their boat at the bathing place and to dine at the château. They in turn invited the princess to come for a row, little thinking she would accept, but to their surprise she did; as the stream was very strong and the princess's weight (estimates varied from twelve to fourteen stone) brought down the stern to within two inches of the water, Mansfield was relieved when she was safely landed.

Figure 4.1. The crew of the *Water Lily* plan their trip, looking out over the Thames.

The crew ignored the dire warnings about the dangers of attempting to boat down the rocky passages of the Danube especially through the Strudel (rapids) and Wirbel (whirlpools) below Grein. Although white breakers surrounded them on either side, steady pulling and careful steering on both stretches of the river brought the *Water Lily* through without mishap. Indeed the greatest problems they encountered on the Danube were the shallows between the islands which often obliged them to jump out of the boat to avoid holing her bottom.

After shooting beneath the old stone bridge at Straubing, they spent the night at Deggendorf. The town ominously turned out to a man to see them start but they passed the rapids between Vilshofen and Passau without mishap and reached Linz. At Grein they were greeted by the Singing Club which had composed a song in their honour but by now such adulation by the population had become a bit of a bore.

Their voyage not only attracted much local attention but was widely if often inaccurately reported in the press. Of the many different notices of their proceedings which appeared in no less than nine Viennese papers, all were either misinformed or exaggerated. The only accurate one was sent by *The Morning Chronicle*'s foreign correspondent, who reported their departure from the Austrian capital some days later. One reason why the papers wrote up their voyage so extensively was the shortage of news due to the very strict press censorship in Vienna. Spy fever was rampant and even Thomson's headpiece — 'a sporting wideawake' — was occasion for him to be stopped by a plain clothes police agent, informed that such hats were considered revolutionary and that he must return at once to his hotel to change it.

The Log of the Water Lily and *The Water Lily on the Danube* were reprinted five or six times, the fifth edition including details of a less successful cruise made on the Saône and Rhône in 1853. Although first in the field of inland boating literature, the narrative suffers from verbose digressions about places visited, anecdotes, gossip and comment on the idiosyncrasies of foreigners whose habits amused Mansfield but were of perhaps more interest to his contemporaries than they are today.

Edmund Harvey

In 1854 appeared *The Journal of an English Pair-Oar Expedition through Europe*, entitled *Our Cruise in the Undine*, whose author, Edmund Harvey, claimed that his earlier travels had been confined to the canals and rivers of Cambridgeshire. He and two colleagues, referred to as the Professor or the Doctor, started their cruise from Paris and after ascending the Seine, they followed the river Yonne to Laroche where they entered the Burgundy Canal. Here they observed that horses were seldom used to tow the barges. Instead mother on one bank, son on the other, hauled while father steered. Some of the lock-keepers fancied the oarsmen were canal inspectors, which caused them to take a lively interest in the state of the locks and the account books submitted for their scrutiny. This may have been because the Emperor was due to pass along the canal in a few days' time and they were mistaken for the officials to pilot his steamer.

At St Jean de l'Osne they were greeted with intense curiosity by the local people who treated them with great civility and respect. People began to collect from three in the morning to see them start and they left the town 'amidst tumultuous cheering and great waving of handkerchiefs'. From the Saône they passed on to the Canal du Rhône au Rhin.

At Rochefort on the Doubs the crew had to share a bedroom at the village inn with three bargees. The following night they shared the only room at a roadside inn with nine other people. Indeed they rarely spent a comfortable night. At Gernsheim they stayed at a small house by the riverside where the mosquitoes forced them to wear their mackintoshes in bed and smoke their pipes. Morning found them bleary-eyed, their hands and faces covered with blotches and bites.

Mesniers are Inspectors of the Canal.

Figure 4.2. Mistaken for Canal Inspectors, the crew of the *Undine* rise to the occasion.

A minor mishap a few miles past the junction of the Basle & Strasburg Canal nearly brought the excursion to a halt. Because of the noonday heat the *Undine* was being hauled from the towpath when they met an empty fly barge drawn by two horses. The *Undine*'s bowline became entangled with that of the barge and before it could be cut, the stern of the barge had scraped the pair-oar breaking two of her thwarts. A carpenter at Kembs was found to make a new set of elbows and 24 hours later all had been repaired. There then arose a problem over the account. Robert, the carpenter, wanted 15 francs, his companion four, while the Professor offered five in all. The dispute was referred to the village mayor, a plain rustic in sabots, who finished feeding his pigs, washed his hands and sat in judgment. After an hour's palaver, an award of six francs fifty centimes was made to the workmen so, except for the loss of time, the crew was well satisfied with the outcome.

The stretch of the Rhine between Huningen and Strasburg was in flood with a tremendous flow and this proved the most hazardous part of their trip. The navigation of the river, 'cut up by innumerable islands and sand banks necessarily abounds in rapids and falls in water', required considerable skill as well as caution. They nearly capsized several times when the swift current caught the keel of the boat making it impossible to steer her. At times they estimated they were racing along at 15 to 18 miles an hour. Once the boat's stern swung round over the edge of a whirlpool and, should the boat have capsized, the fate of the Doctor, who could not swim, might well have been sealed. 'Now we are sweeping down a rapid so shallow that we can see the bottom, and rush past the trunk of an old tree, which itself seems as if it was being pulled up against the stream by some invisible power; now we reach the end of the shallow, and the water is boiling and bubbling about us in all directions, and the sharp crested waves come rattling against the side of the boat like so many undertakers at work.'

The 70 miles from Neuenburg to Kehl was covered in six hours and a quarter and the Doctor doubted if even the *Water Lily* on the Danube had met with such a continuous rapid stream.

At Mannheim the *Undine* was stowed in the same stable used by the *Water Lily*. After visiting Heidelberg where they witnessed a duel, the voyage down to Cologne proved uneventful and rather than continue down the Rhine to Rotterdam, they placed the *Undine* in a railway truck full of wool sacks and sent her to Ghent whence they rowed to Bruges and Ostend. Altogether a more interesting and eventful voyage than that of the *Water Lily*, but ten years and more were to pass before another literary account of a continental excursion was to appear.

John MacGregor

It was not, indeed, until the advent of canoeing that boating became relatively inexpensive. The rapid rise in its popularity was largely due to John MacGregor, who designed his own *Rob Roy* canoe and described his exploits in a series of narratives which won wide public acclaim. A thousand-mile canoe voyage on the continent in 1865 was followed by one in Sweden the following year and another on the river Jordan in 1868 when he was actually captured in his canoe by Arabs.

The first *Rob Roy* canoe built by Searle's of Lambeth was 15 feet in length, 28 inches in breadth and nine inches in depth. Her weight was 80 pounds and she drew three inches of water. Besides a twin-bladed paddle seven feet long, she had a mast, lug sail and jib; a silk Union Jack fluttered at the prow and the name *Rob Roy* was painted in blue letters on the stern.

The first European voyage MacGregor planned led over mountains, through forests, across plains and involved nagivating no less than 11 rivers, including the Rhine, the Danube, the Reuss and the Marne, venturing upon six lakes and touching upon six canals in Belgium and France. Being a fair draughtsman, MacGregor, sketchbook in boat, illustrated his exploits himself.

A Thousand Miles in the Rob Roy Canoe on Twenty Lakes and Rivers of Europe, with woodcuts from the author's drawings, was published in January 1866. Within a month a new edition was printed. *The Times* gave the book two and a quarter columns of unmixed praise and the second edition of 2,000 copies was sold out in five days. The Emperor Napoleon III read it and decreed an exhibition of pleasure-boats in 1867. MacGregor promptly sent him a specially bound copy and a photograph. By the end of the year, 8,000 copies had been sold.

MacGregor's popularity lay in his appeal to the adventurous Englishman who wanted to be a traveller rather than a tourist. As early as 1865, he protested against the packaged tour and the ardent followers of Baedeker: 'Year after year it is enough of excitement to some tourists to be shifted in squads from town to town, according

Figure 4.3. John MacGregor in his sailing canoe on Lake Zug.

to the routine of an excursion ticket. Those who are a little more advanced will venture to devise a tour from the mazy pages of Bradshaw, and with portmanteau and bag, and hat-box and sticks, they find more than enough of judgment and tact is needed when they arrive in a night train, and must fix on an omnibus in a strange town. Safe at last in the bedroom of the hotel, they cannot but exclaim with satisfaction "Well, here we are all right at last!"'

The churches and galleries, mountains and caves, ruins and battlefields had by now been pretty well seen. The rivers and streams at home and on the continent, however, were scarcely known to the English tourist and the enjoyment of such treasures was enhanced by the energy and pluck required to get at them.

Formation of the Canoe Club (1866)

The enthusiasm which greeted the appearance of MacGregor's book caused him to found the Canoe Club. The first meeting was held at *The Star and Garter* at Putney on 27 July 1866 when 20 members were enrolled and MacGregor elected captain. Any gentleman who owned a canoe or hired one by the year could be proposed and elected a member — 'one blackball in five to exclude'. The first list of members included distinguished oarsmen, travellers, alpine climbers and athletes, and within a few months the club was patronised by H.R.H. The Prince of Wales who accepted the invitation to become the first commodore.

The minutes of the first meeting of the Canoe Club announced that voyages were already planned for the summer of 1866 for England, Wales, Scotland, Ireland, Norway, Sweden, Denmark etc. Certainly the year saw the *Rattlesnake* launched in the Trent, the *Robin Hood* being paddled to Windermere, the *Rapid* and the *Rover* starting for Scotland while the *Ripple* set off for the Rhine. The *Rambler* was touring 'promiscuously' while it was hoped that the *Reverie, Ranger* and *Romp* would soon float gaily. (It was a club rule that all names should begin with the letter 'R'.) Several members, it appeared, were also intending to meet with their canoes and dogs on the Clyde. The logs of club members contained details of many amusing adventures as well as three or four upsets and 'canoe catastrophes' but only a single life was lost — that of a dog. One member bumped into a bridge on the Severn and was left clinging to it while his boat departed bottom upwards. Another went over a weir sideways, the occupant upset and the canoe broken to pieces. A third struck on a lasher and was capsized twice in two days while another coming in on a rough sea was seized by a crowd of compassionate fish wives, watching for him on the shore, who carried the man and his boat in their brawny arms, rescuing him as though he had protested 'I will be drowned; nobody shall save me'. To all of which reports *The Pall Mall Gazette* commented that the Canoe Club 'unites the maximum of danger and discomfort with the minimum of utility' which caused MacGregor to urge the writer to join their club and recover his good humour.

There were strange if not wholly unexpected reactions from some Victorians. One who had just spent his vacation in Brighton asked MacGregor 'was it not a great waste of time?', and another said: 'Don't you think it would have been more commodious to have had an attendant with you to look after luggage and things?'. Membership of the club increased and many voyages took place in 1866. One rowing blue (the Hon. James Gordon) sailed his canoe across the Channel to Boulogne and paddled through France to the Mediterranean at Marseilles and back via Genoa through the Italian and Swiss lakes, the Meuss and the Rhine. A second paddled across the Channel to France, a third went down the Danube, the Moldau and the Elbe. A fourth carried a dog on board which on shore dragged the canoe on wheels. Three went round the rocky coast of Skye of whom two crossed to the island of Rhum. Others cruised the Clyde, the Thames, and the Irish streams while one took his boat to India and another to Australia.

P. G. Hamerton

Amongst the Canoe Club's early members was P. G. Hamerton, the art critic and writer, who designed a paper canoe, and in 1867 explored the river Arroux, a tributary of the Loire. The tale of this adventure was recounted in *The Fortnightly Review* and later illustrated with 37 etchings in book form under the title *The Unknown River*.

Hamerton explained his enjoyment of venturing on a nameless European river as being the assurance of 'finding much that is worth finding, and of enjoying many of the sensations which . . . give interest to more famous explorations', and added that it was necessary 'to the complete enjoyment of an excursion of discovery, that the region to be explored, whether mountain or river . . . should not have been already explored by others, or at any rate not with the same objects and intentions . . . a traveller who is fond of boating has an especial pleasure in descending some stream of which it may be safely presumed that nobody ever descended it in a boat'.

Warington Baden-Powell

Another enthusiast was Warington Baden-Powell, elder brother of Robert, who is credited with the introduction of sailing canoes and whose voyage in 1870 to the Baltic and across Sweden was recounted in *Canoe Travelling* which also included practical hints on building and fitting up canoes.

The reader's main impression is of two youthful canoeists soaked to the skin, braving miserable weather for much of their six-week cruise. Drenching rain, squalls, head winds and rough water must have made MacGregor's voyages seem more like luxury cruises.

Their food and accommodation was also often frugal. The second night was spent on the rocky banks of the Gotha river inside the canoes lying on a rug in their wet clothes covered only by a mackintosh. On occasions when searching for somewhere dry to rest for the night they found no house, no sign of human life, darkness increasing and rain threatening, drizzling and pouring. When they did find accommodation, it was often no more than a barn or hay loft. Even at the Grand Hotel at Oxlö, they could dine only on black bread and butter, eggs, milk and hot water and share a large upper chamber serving as a dormitory. 'H', whoever he was, does not appear to have said a word in complaint and it is intriguing that Warington should have made no references to his travelling companion throughout the book. Baden-Powell's account was followed by what was now traditional advice on stores, clothing and cooking. Tea was recommended as being less bulky than coffee, Leibig's beef essence as useful when fish and eggs were not available, brandy to purify bad water and to drink when no beer or milk was to be had. Indispensable items were gold, tool chest, varnish, putty, sheath knives, fishing tackle, breech-loading pistol and a 'dark eye-glass for use when steering with the sun in front'. An Inverness cape was recommended for chilly nights and a mackintosh sheet to form a tent. The 'Rob Roy' cooker on its tripod was described as a 'little wonder', able at times to boil a quart of soup in five minutes but unruly in wind.

Shortly after *Canoe Travelling* had appeared in 1871, John MacGregor dined with Baden-Powell. There was an age gap of 22 years and one can only speculate how the two men got on. However Baden-Powell found himself differing with MacGregor in one respect when he wrote 'To some men there is an intense enjoyment in being alone with their thoughts in a foreign country for months together, the spell only to be broken at intervals by the necessary intercourse with the natives. I have enjoyed many a short solitary cruise, solitary because mine was the only canoe in those parts of the world, but on a long trip I should prefer as many companions as possible, for even when two are together every trouble is halved and pleasure doubled'. And yet on his voyage to the Baltic there is not a single mention of any singular incident or conversation with his companion 'H'.

James Molloy

By the early 1870's boating excursions 'had' according to Mansfield 'become so popular that scarcely a year passes without one or more boats being conveyed across the Channel to investigate the Continental streams' Of many such expeditions no record exists. Several, however, were written up and published. Amongst these are J. L. Molloy's *Our Autumn Holiday on French Rivers*, a tale not unlike *The Log of the Water Lily* but full of incident, written with more humour and well illustrated by Linley Sambourne who was to become better known for his work with *Punch* and his illustrations for Charles Kingsley's *Water Babies*.

In this holiday adventure a 40-ft. four-oared outrigger crewed by oarsmen of some ability was rowed up the Seine from Caudebec, down the Loire and up the Breton Canal. It was a voyage of many incidents, some of an unfortunate character. On the second day the *Marie* was caught in rough water above Duclair and sank, the crew being saved from drowning by the timely appearance of a local ferryman. Resuming their voyage a week later, the boat once again nearly foundered in stormy weather and had to be shipped some 80 miles upstream to Vernon. At St Mamès they discovered the Canals du Loing and d'Orléans closed for repair and on the Loire near Tours they were fortunate to survive the numerous sandbanks and quicksands. On the Breton Canal they had a minor mishap, running into a lock wall in the dark, but the damage to the boat was soon repaired.

Figure 4.4. A tricky embarkation for the crew of the *Marie*.

Figure 4.5. The *Marie* at Samois.

Both the weather and their accommodation varied from excellent to atrocious; some hotels were full, some were outrageously expensive; others could not be bettered for their excellent tables and willing service; the tiny country inns provided very modest beds but cheap fine fare and local wine. 'Hotel homely and comfortable as one could wish, and a very pretty landlady, who popped down ducks and partridges the moment we entered. Came across some excellent wine called Beaugency', ran one comment. The voyage was rated a success; the author, commenting that 'from the first to the last stroke of the cruise, the interest went on increasing . . . and two months more of such wandering would have been gladly welcome'.

Molloy's account of the rowing exploits of his colleagues was the first narrative to dwell on the humorous side of boating. His predecessors had rather too faithfully logged the details and omitted the fun, and it is to Molloy that both Stephenson and Jerome owe a debt.

William Moens

A cruise of quite a different character occurred in 1875. William Moens, the son of a wealthy Dutch merchant who had emigrated to England, decided rather on the spur of the moment to take his 45-ton steam yacht *Ytene* up the Seine and through the canals of France and Belgium. Already Moens had had an eventful life. In 1865 while travelling with his young wife in Italy, he and a friend had been captured by a band of brigands. A ransom of £8,000 was demanded and Moens remained a hostage for four months, being forced to sleep in caves and scramble across rugged mountainous country while Italian soldiers frequently harried the band from a distance. Eventually released on payment of £5,100, he wrote a lively two-volume account of this episode in *English Travellers and Italian Brigands*.

In 1869 Moens had sailed his steam yacht *Cicada* from Lymington up the Rhine to Strasburg and by the French canals to Paris and Le Havre passing 219 locks and five tunnels. Moens recalled that his voyage was regarded with great interest in France as no steamers had successfully traversed the route before although two steam barges had failed in the attempt. Although he apparently kept a detailed log of this trip — recording, for instance, the fact that he passed the 2½-mile Mauvages tunnel in 37 minutes — it was never published.

It was therefore perhaps a little surprising that he should decide to write up his account of a similar voyage six years later. There is no doubt he was a man who loved detail. Besides descriptions of visits to places of historical, ecclesiastical and industrial interest, Moens itemises the amount of coal burnt, oil used, canal dues, pilots' fees, and even, the price of peaches. As one of the barge masters who met Robert Louis Stevenson reported, 'He came ashore at all the locks and asked the name of the villages, whether from boatmen or lock-keepers; and then he wrote them down. Oh, he wrote enormously! I suppose it was a wager'. An original reason for taking notes, thought R.L.S.

The crew consisted of Moens as captain and chief engineer, Fisher the engineer and stoker, Miller the mate who was in charge of cabin duties, and Allen the boy cook. Guests were 'received' on board from time to time and taken on excursions, sometimes six or seven, but little mention is made of them or of his wife Anne except when he asked her to comfort the women and children huddled with their bundles of bedding on the canal banks after being evacuated from a sinking coal barge. Sadly her name which appears in print abbreviated as 'A' occurs only three times in the book while details of lengths and widths and distances, and the suppliers' quality and price of beetroot and steam coal, are regularly provided. There were also two black and tan toy terriers on board, Titus and Lill, who took little part in the events to be narrated.

The voyage across the Channel by night was uneventful and little of interest is recorded on the journey up the Seine except of the need for pilots due to the hazards in passing the bridges destroyed during the Franco–Prussian War (1870–71). At Rouen they hired Rousillon, a retired chief pilot, 'a quiet obliging man who kept a wine shop and billiard table at 20 Quai de Paris'. Watching the haymakers at St Oissel twist the dry grass into small bundles reminded Moens of his experiences of Italian mountain life 10 years before.

A couple of days were spent admiring the beauties of Les Andelys where Moens brought out his photographic apparatus. At La Roche Guyon they found the London steamer *Arion*, bound for Paris, stopped and unloading cargo into barges as the water level had fallen too low — she was drawing five feet. After a brief stop at Mantes, they moored for the night above Meulan and reached Paris the next day where they berthed opposite the Tuileries gardens.

After nearly a fortnight in Paris visiting among other things the Maritime Exhibition in the Champs Elysées and the Geographical Exhibition in the Louvre, and during which Fisher, who was a great tram traveller but knew no French, found himself at the wrong terminus of a tram route and had an unpleasant night out, they steamed down the Seine past Bougival to the river Oise.

Five miles up the Oise they observed Chateau Vauréal then passing between Erangy and Pontoise, they proceeded past Champagne and Royaument. 'All the barges met were coal-laden, coming from the Belgian coalfields; they were very deep, having two hundred and seventy tons on board, and drawing six feet of water.'

Moens is keenly interested in all he sees, noting a nail factory, a sugar refinery, stone foundries, and ironworks as well as discoursing on churches and buildings of historic interest. Passing under triple-arched Pont St Maxence, built in 1774, he comments that this was one of Perronet's best works. They saw the pleasure boats on the lakes of the Palace at Compiègne used in the time of the Emperor. 'Merry and gay was the forest then; parties and sports of every kind made it ring with joy and laughter, all of which the poor residents in the forest villages now look on as things of the past, and not likely to return except with a return of the Empire.'

After leaving Compiègne on 27 September with a strong wind blowing behind them, they passed Jonville and Chauny locks which were so crowded that they were stowed fender to fender with the barges. They procured a carriage to view the dungeons of Courcy but were refused permission to view the glass works at Chauny by the managing director Monsieur Bivet, who was really most 'uncivil'. Moens trusted that 'he, when visiting England, will never meet with the reception that he gave an English gentleman and lady who wished to see foreign industries'. The factory employed some 2,000 hands working in shifts day and night and the nearby chemical works another 1,500. Moens couldn't resist adding, 'we heard that the workmen suffer much in their health in these works, many losing all their teeth'.

The following day they passed ten or so locks on the St Quentin Canal, noting the tramways to the coalmines near Voyaux, and that the people in this part were poorer and the houses more wretched in appearance. After passing the junction to La Fère, commercial traffic diminished. At St Quentin, Moens noticed that a baker was delivering bread in his cart and still using for each house a wooden tally, on which a notch was cut for each loaf.

Two hours after leaving St Quentin they passed through Tronquay tunnel whose centre height (17 feet) was just two feet higher than their funnel. On the other side they were forced to lie up behind a long line of barges which were being towed by a submerged chain. Ahead lay the long tunnel of Riqueval and further delay ensued when the line heaved-to to allow a long string of barges to pass. At about 9 p.m. there was quite a commotion when it was learnt that one of the oncoming barges heavily laden with coal was sinking in the tunnel. 'This was a pretty state of things, and we soon thought that our route to Belgium would be barred for weeks, and that we might have to retrace our way back again. The tug steamer soon, however, emerged from the arch, and came to a standstill when three or four barges were out of the tunnel. It was the first that was injured, and she was already sunk in the water to within three inches of the gunwale.'

It took 70 minutes for the *Ytene* to pass the tunnel. After entering the Scheldt Navigation at Cambrai they cruised through Belgium where the vessel created rather more interest since it was the first steamer ever seen. After spending a week in Brussels they steamed down to Calais and back to Lymington having covered 1,115 miles in two months and burnt 15 tons of coal.

A rich man's rather uneventful voyage had been accomplished with little of the wealth of human experience which Stevenson was to enjoy when making a similar journey by canoe the following year.

R. L. Stevenson

It is said that Robert Louis Stevenson's *An Inland Voyage* was inspired by Molloy's account, but Stevenson's familiarity with MacGregor, Hamerton and Moens suggests that he only awaited the opportunity to combine boating with writing a book. His canoe journey with Sir Walter Simpson in the *Arethusa* and *Cigarette* through the canals of Belgium and France in 1874 is more remembered for its vivid prose style than its eventfulness, but poor weather made the cruise rather a cheerless experience, as those who go boating in Northern Europe have often discovered. However, Stevenson was not the first man to endure the hardships of foreign travel with the object of making a book out of it and some 'coin'. Ten days after their departure, he wrote from Compiègne 'We have had deplorable weather quite steady ever since the start; not one day without heavy showers; and generally much wind and cold wind . . . I must say it has sometimes required a stout heart; and sometimes one could not help sympathising inwardly with the French folk who hold up their hands in astonishment over our pleasure journey. Indeed I do not know that I would have stuck to it as I have done, if it had not been for professional purposes; for an easy book may be written and sold, with mighty little brains about it, where the journey is of a certain serious-ness and can be named. I mean, a book about a journey from York to London must be clever; a book about the Caucasus may be what you will. Now I mean to make this journey at least a curious one; it won't be finished these vacations.'

Stevenson and Simpson travelled without tents and lodged where they could, with mixed results since R.L.S.'s unkempt appearance and unorthodox clothes held against him when it came to obtaining lodgings. *An Inland Voyage* is full of references to their nightly stops. The *Hotel de la Navigation* was the worst feature of Boom, boasting dismal parlours with sanded floors, an empty birdcage and food of a nondescript character. Mauberge had a very good inn, *Le Grand Cerf*, but at Pont sur Sambre, where the landlords and tavern keepers took them to be pedlars, they were fortunate to secure a twin-bedded room in the loft, furnished with three hat pegs and a table. There was not so much as a glass of water and in the morning they found two pails of water behind the street door for their ablutions. At Landrecies the pair found better accommodation with 'real water jugs' and plenty of furniture, and at the *Golden Sheep* in Moy they found excellent entertainment and partridge and cabbage well above expectation; however at La Fère the landlady took one look at their bedraggled appearance and their limp india-rubber bags and told them to find beds in the suburbs. 'We are too busy for the like of you.'

An Inland Voyage remains the finest literary narrative of a 19th-century boating expedition. Stevenson recorded the sense of freedom to be found in canoeing, the appalling wretchedness of continuous wet weather and the problem of finding food and accommodation in out-of-the-way villages. He considered that of all the creations of commercial enterprise a canal barge was by far the most delightful; he projected an old age on the canals of Europe. 'It was to be the most leisurely of progresses; now on a swift river at the tail of a steamboat; now waiting horses for days together on some inconsiderable junction.' R.L.S. even bought a barge and christened her *The Eleven Thousand Virgins of Cologne* but ill health prevented the realisation of his dream.

P. G. Hamerton on the Saône

In 1886 Hamerton, now in his fifties undertook quite a different sort of voyage to his earlier one when, with the young American artist Joseph Pennell, he set off to write and illustrate a book on the Saône.

Figure 4.6. A Saône steam tug, drawn by Joseph Pennell.

Hamerton hired an 80-ft. long Berri barge the *Boussemroum,* which had a flat bottom and whose hull was divided into compartments which normally held the cargo as well as a donkey house and hay store. This area he converted into a sort of houseboat by the erection of a camp inside it. Each tent had an iron bedstead, camp chair, paper ewer and basin, chest, table and Japanese matting on the oak floor. A quarter deck with an awning and camp stools provided an area for viewing and relaxation. The largest tent on the deck formed the saloon which served as dining room and library and held the sole armchair. Here when the weather was bad Hamerton wrote in comfort while Pennell did his best to sketch.

Some 40 years later Pennell devoted a chapter of his autobiography to his voyage on the Saône. It is an interesting account and confirms what many readers of *The Saône* might have guessed; that the adventures Hamerton recorded were from his point of view but 'scarce from' Pennell's. Hamerton had his cabin and his books and his lamp, but Pennell had a tent, and he was not used to tents, particularly ones which made a sort of basin which caught the rain and which overflowed when the occupier turned in his sleep! Furthermore although Hamerton was 'delightful and kind, there was something about him – he did not just come off. Everything with him was like the *Boussemroom*, arranged according to his theory. He loved camping, and sailing, and art too. But there was through it all a something, a want of humour, of fun, a deadly seriousness that was his great enemy and barrier and defect.' Pennell and Kornprobst would spend night after night in Hamerton's cabin in serious talk, and 'sometimes there was even a petit verre de kirsch which we bought from the lock-keepers who made it. But as we left his cabin, the Captain would whisper as we made our way back to our windy tents – "allons au café", and after a brandy would say "Il est charmant, mais, mon cher, quel type !".'

The *Boussemroum* presented one big drawback: it was almost unmanoeuvrable in any sort of a wind. Ordinarily it was loaded with coal so that its hull was well down in the water, but without this ballast it rose six feet. Buffeted by the slightest breeze, it was all the poor donkey could do to avoid being pulled into the river and it was necessary for the pilot to push constantly with a pole to keep the keel from striking the bank. On one occasion near Corré the barge grounded every minute or so with a shock which made writing or drawing impossible.

The voyage held one moment of suspense. Relations between France and Prussia had continued to be strained during the 1880s and it was perhaps unfortunate that at the time of Hamerton's voyage France was in another grip of spy fever. At Pontailler the barge was boarded by four gendarmes (one of whom was remarkably aggressive), the travellers were arrested and their drawings confiscated. Only prompt action by Mrs. Hamerton secured their release and the return of the drawings the following day.

At Chalon, Pennell decided he had had enough of living on the *Boussemroum* and planned to complete the voyage by public steamer. This intention was thwarted by his failure to obtain official permission to sketch at Lyons and rather than abandon the book he agreed to make drawings from the author's sketches, so Hamerton resumed the voyage in his 24-ft. steel-hulled catamaran, *L'Arar*. With him sailed his younger son Richard and nephew Michael Pelletier.

Hamerton's mode of travelling must have been something of a strain for the younger men. Every member of the expedition had his appointed duties, and was expected to perform these with the strictest regularity. His wife recalled that he was a man who liked to 'marcher militairement'. They rose at five, breakfasted at seven, lunched at one, dined at seven and retired at ten. 'Everything has its place [in the boat] according to two systems, one called "night order" the other "day order", and each of us knows the place of each detail under the other two. There is consequently no confusion, and I [Hamerton] am spared the trouble of giving directions.' His style of living was founded on the principle of decency without luxury. *Vin ordinaire* was the rule and good Burgundy was reserved for 'state occasions'. Hamerton had no desire to have women or children on board: children because they would be in the way and make a noise, and 'I should like to know the woman personally before intrusting my peace to her for many weeks. She might possibly be a talker, or even a scold, she might be dirty and slovenly in her habits, all which would spoil the pleasure of my trip completely'.

The Saône: A Summer Voyage appeared in 1887. The reviews were, according to Mrs. Hamerton, 'hearty' and *The Spectator* described the story as 'delightful throughout'. It did not, however, sell well and was never popular.

Davies, Doughty and Jerome

G. Christopher Davies was a noted yachtsman, photographer, angler and naturalist who shared with Henry Doughty a love and knowledge of the Norfolk Broads. Davies, however, made only one continental cruise and his account of sailing on the rivers and canals of Holland and the north of Belgium was not done in a wherry, because of the cost of the tow across the North Sea, but in a 60-ft. vessel drawing five feet of water, which proved a trifle too great for the smaller canals.

Henry M. Doughty on the other hand voyaged abroad with his family in the Norfolk wherry *Gipsy*. Fifty-three feet long, it had the comfort of a Thames houseboat, was

Figure 4.7. Miss Doughty's record of the crew of the *Gipsy*.

safe, fast and handy under sail and could float in three feet of water. The vessel was towed across the North Sea and cruised the Friesland Meres and the Netherlands. After berthing *Gipsy* for the winter in Friesland, a second voyage in 1891 took them up the Elbe and through the Mecklenburg Lakes to Bohemia and Dresden. In the same year American explorer and writer Poultney Bigelow with two novices paddled canoes down the Danube. It was a well-publicised trip but in fact only the novices actually reached the Black Sea.

All 19th-century boating narratives had common features. The crew remained anonymous even if the pseudonyms were thinly disguised; the types of craft and their stores were fully tallied; the boats were given names and their logs gave details of the food and drink consumed, the quality of accommodation found and the reception given to the adventurer as he arrived at each new place. It was left to Jerome K. Jerome in *Three Men in a Boat* to continue this tradition and to produce the most read and possibly the best humorous account of a boating expedition.

Early Twentieth-century Adventurers

Boating literature of the first decade of the 20th century reflects the continued interest and growing number of people able to afford both time and money to cruise on the continent. Nearly everyone kept a log for the amusement of the family and friends; those that described some newly 'discovered' waterway were published in *The Field* or *National Geographic Magazine*. The best often appeared in book form although they form a list of varying literary value.

Before the Great War suspended pleasure boating in Europe, Wykehamist Roger Anderson had made three cruises through Sweden (1907), Norway and Sweden (1908) and Finland, Russia and Sweden (1909) by sailing canoe and dinghy. In *Across Europe in a Motor Boat*, Henry Rowland chronicled the adventures of the crew of the *Beaver* as they cruised nearly seven thousand miles through Europe by way of the Seine, Rhine, Danube, only to meet shipwreck in the Black Sea.

Novelist Arnold Bennett was a keen yachtsman who described in *From the Log of the Velsa* his trips by motor yacht through the Norfolk Broads, Belgium, Holland and the Baltic. But it is artist and countryside explorer Donald Maxwell who provided more original accounts of boating adventures. As a young man of 24 he surprised the inhabitants of the Swiss village where he was studying by having a flat-bottomed yacht built although the nearest navigable water was 20 miles away. The *Griffin* had to be carted over the Riken Pass to Lake Zurich before she could begin the journey back to Teddington, not solely under her own sail, as from time to time she travelled by train and barge, before being lifted on to a freighter for crossing the North Sea. The book, *The Log of the Griffin*, was well reviewed, *The Dundee Advertiser* critic considering that there had not been a book so fresh, amusing and funny since *An Inland Voyage*, and *The Leeds Mercury* noting that in his method of writing Maxwell 'reminds us of Mr. Jerome at his best'.

Four years later Maxwell undertook a more adventurous voyage with a colleague from the coast of Holland to Sulina on the Black Sea, later described in his *A Cruise Across Europe*, graced by his own drawings. Although slightly delayed by being arrested at Dordrecht as suspected spies, and on the Rhine by adverse winds, the difficult part of the cruise was passing through the Ludwig's Canal. 'Nearly a week elapsed before the *Walrus* had obtained official permission to pass through the canal, the property of the Bavarian State. She had to be carefully measured, and numerous descriptive entries were made on a paper forwarded to Nürnburg. In some places, especially where it bridges a river or ravine, the channel is only about three feet in

Figure 4.8. The *Griffin* in the Alps.

depth. This probably accounts for the exhaustive examination which every barge under-goes before entering the first lock.' They entered the first of the canal's 100 locks on 21 March 1905.

There was little traffic on the waterway and less than ten working craft were encountered. The scenery recalled familiar river scenery in England and 'if the reader can imagine that the Wey and Arun Canal had lost itself upon a spur of the Alps, he would get some idea of its character'. The *Walrus* weighed 30 hundredweight and more, and the two travellers not surprisingly found towing her most of one hundred miles hard work. However, sailing along the summit level brought some respite and the sight of a sail was a phenomenon to the inhabitants of the villages below. 'Greetings and good wishes were shouted to the man in the *Walrus*, but here as at other places the man on the tow-path was taken for a hired slave, and treated accordingly.' Follow-ing an easy passage through the 'Iron Gates' below Budapest, the trip culminated in anti-climax as Maxwell caught malaria and was confined to bed by the time they reached their destination.

Between the Wars (1918–1939)

The problems of Europe following the aftermath of the Great War prevented an immediate return to pleasure cruising, but in the early 1920s Leslie Richardson had a 26-ft. motor boat built by Rondet of Nantes, with a 16-hp. engine and auxiliary sails, in which he cruised between Brittany and the Riviera. A more interesting venture, however, was American Negley Farson's voyage in a motor yacht with his wife from the North Sea to the Black Sea in 1928, described in *Sailing Across Europe*. They were towed by the 'Kette-Boot' which moved on a continuous chain 190 miles up the Main to Bamberg where 'a fat little man' who was guardian of the Ludwig's Canal remembered that perhaps in the past 20 years only two fine pleasure boats had passed through. Like Donald Maxwell, Farson was forced to bank-haul the *Flame* through much of the canal because of the weed, and wrote 'in the old days the weeds were cut, and one can still see the scows. But now the canal is almost deserted — a veritable inland Sargasso Sea'.

C. S. Forester and the Outboard Motor

Not many people are aware that the spinner of the stirring seafaring adventures of Captain Hornblower and author of *Brown on Resolution* wrote two delightful accounts of his boating experiences in Europe with his wife Kathleen towards the end of the 1920s — the first accounts of voyages relying on an outboard motor. His novels might give the impression that C. S. Forester was more of an ocean-going yachtsman than a boat camper but he wrote, 'There are I know people who wrinkle intolerant noses at the thought of rivers, and who insist that there is no joy in the whole world save the sea. Of course they are wrong, although I will go so far as to say that the sea is a possible second best, and that beating to windward in a five-tonner has its points'.

For the author and poet it was rivers every time. 'The swirl and eddy at a deep corner, the sight of grey reaches of river in summer rain, the wooded hills that climb from one bank and the fat meadows which stretch from the other — the very writing about these things starts an ache of longing in one's breast.' Forester confessed that he had a passion for running water — 'I have tugged at oars and pulled paddles, and pushed punt poles on odd rivers all over the place. I have swum for my life . . . I have blistered my hands . . . I have suffered torments of pins and needles in a Rob Roy . . . Most powerful influence of all, I have a wife with the same craze rather more developed who has shared most of these experiences.'

Figure 4.9. Donald Maxwell steering the *Walrus* on the Ludwig's Canal.

After rejecting cabin cruisers as being too confined for a long voyage, skiffs and punts as too slow and unwieldy, they chose a camping boat – a punt-like dinghy with a five-foot bulge near the stern, drawing but four inches of water, and with a freeboard of 15 inches. Three hoops bore a camping cover of Willesden canvas with a headroom of five feet two inches. Two canoe paddles were often but not always assisted by an Evinrude 4-hp. outboard, a motor whose initial idiosyncracies dominated their lives until its owner learnt that in the operating instructions 'not running satisfactorily' really meant 'not running' and that an extra half turn of a needle valve was the solution. Once this discovery had been made 'we found out why I had torn the skin from my fingers unavailingly, why we had made exhibitions of ourselves before the military orphans, why our whole voyage so far had been tainted with doubt as to our ability to move on next day, why we had had to fight so frantically with our paddles on occasions instead of having the motor to do our work for us'.

The *Annie Marble*, named after a character in one of Forester's early novels, was shipped to Le Havre in the summer of 1928 and entrained to Rouen whence she began a three-month voyage up the Seine and the canals du Loing and d'Orléans to the Loire and down to Nantes. In spite of variable weather and occasional frustrations it was a very happy adventure.

The incidents of the first voyage were generally unexceptional although on the Canal d'Orléans near Lorris they struck the bank at full speed and broke off a propeller blade. Only the lock-keeper and his wife's hospitality made their enforced stay of nearly a week less of a trial while they awaited the arrival of Forester's sister, bringing a new propeller.

The bargees on the Seine who towed the boat upstream when the engine failed, and nearly everyone else they met, were both generous and helpful. The sandbanks and shoals and the many arched bridges on the Loire were constant navigational hazards but the *Annie Marble* was the ideal vessel for such a voyage. Boats of any type were rarely seen and the villagers were 'frightfully puzzled by their presence'. At Blois, Forester's sister returned to England leaving C.S.F. and Kathleeen to what the author admitted was pure happiness as they floated down the Loire to Nantes in perfect weather. 'Nothing much happened, and yet every minute was delight.' Their happiness on the Loire was later to be mirrored by Horatio Hornblower who, when making the descent of the Loire in *Flying Colours*, 'was happier at this time than a life of action and hardship had ever allowed'. Forester went on to admit that while he was writing these passages 'there was a momentary temptation to prolong the voyage; certainly there was a faint regret that the Loire was not as long as the Amazon'.

The Voyage of the Annie Marble being the Story of a Cruise through France with an Outboard Motor Boat appeared in 1929, price 8s. 6d. *The Daily Telegraph* wrote that it was 'one of the most charming books of travel that has appeared these many years'. *The Tatler, Spectator* and *New Statesman* were equally enthusiastic.

The second voyage, in 1929, described in *The Annie Marble in Germany*, was perhaps a trifle less enjoyable due to patches of indifferent weather and the dreariness of parts of the Elbe. From Hamburg they boated up to Magdeburg, then through the Ihler and Planer canals to Berlin and back down the Havel to the Elbe and Domitz. Then by the canalised Elde they passed through desolate heathland to Schwerin and the Mecklenburg Lakes. Rejoining the Havel as far as the Voss canal, a transit of the Hohenzollern canal took them to the Oder and the port of Stettin, whence the *Annie Marble* was shipped back to London.

Except for a new outboard, the vessel was unaltered. After Forester's criticism of the Evinrude, and the author was at pains to explain that his decision to change engines bore no reflection on the merits of that company's products, it was not surprising 'that as soon as the Elto people heard I was proposing to take one of their engines to Germany . . . they insisted on instructing me in the handling of the thing. Possibly — I think it extremely likely — they had heard some rumour respecting my mechanical gifts, and did not feel exceedingly pleased at having their reputation handed over to the mischances of such a hopeless engineer as myself. Anyway, they sent a mechanic down to me who put me thoroughly through the secrets of the engine. But I was so lost in admiration of the mechanic himself, of the way in which he whipped sparking plugs out in the twinkling of an eye, and the familiarity and ease with which he discourse on such abstruse subjects as carburettors, revs., and reduction gears, that I quite failed to take in a word he said. Then they loaded me up with a bag of spares, and (no doubt with the gravest misgivings), wished me a successful journey.'

Major Raven-Hart

The popularity of canoeing during the period between the two World Wars was enhanced at home by William Bliss and on the continent by the literary works of Melville Chater, an American, and Major Raven-Hart, who like MacGregor, ranged far and wide.

Rowland Raven-Hart had found himself at the close of the Great War with impaired health, precious little money and a burning desire to see something of Europe. It was not however until 1930 that he began to canoe across Europe, and it was 1935 before the first account of his explorations appeared in *Canoe Errant*. That winter he canoed down the Nile from Wadi Halfa in the Sudan to Cairo and the following year he cruised the Mississippi from Hannibal to the Gulf of Mexico. In 1938 he paddled down the Arun from Pulborough to Littlehampton before tackling the Irrawaddy from Myitkyina to Mandalay. Finally he tackled the southern hemisphere, but *Canoe in Australia* was not published until 1948.

It is only fitting that this chapter should be concluded by Merlin Minshall's claim to be the first Englishman to sail across Europe. Employing a motorised Dutch sailing barge, in 1937 he voyaged from Le Harvre up the Seine to Paris, then on to Strasburg and through King Ludwig's canal to the Danube. His claim, of course, rests on his choice of the Seine rather than the Rhine as a starting point. The only other west–east crossing, that described by Rowland in *Across Europe*, had been made in an English boat — but one bought and crewed by Americans.

*   *   *

The post-war era has seen no lessening in the desire to explore waterways at home and abroad, and this continuing interest has led Tom Rolt, Roger Pilkington and others to write topical and sometimes amusing accounts of their own wanderings. However, it is a sad fact that many modernised waterways lack the charm of their predecessors.

King Ludwig's canal was closed in 1951 and its new counterpart, if and when opened, will be a very different type of navigation. The large locks on the Main which Maxwell passed have been quadrupled in size. The Iron Gates, and the Strudel and Wirbel of the Danube, have ceased to present a challenge. Many smaller canals have been abandoned and even the excitement of exploring lesser-known rivers and streams may be reduced by the knowledge that the 'degree of difficulty' has been carefully assessed by Michelin and the canoeing and touring clubs. The boating enthusiast finds too, that his geographical area for exploration is now more circumscribed. Voyages in eastern Europe, the Middle East and even Africa are unlikely to be welcomed by the authorities and it is the practical difficulties and frustrations which are now so much greater than the natural obstacles to be overcome. This realisation is unlikely to diminish our desire to go on finding and exploring rustic waterways, but it does rather make us envious of earlier adventurers.

APPENDIX TO CHAPTER 4:

A bibliography of European Cruising, 1833–1939

Compiled by Dr. Mark Baldwin

This appendix lists accounts in English of voyages made on the inland waterways of mainland Europe in privately-owned boats, first published in or before 1939. It is not restricted to books published in Britain, and also includes a few periodical articles. Handbooks largely comprising cruising guides and advice are omitted, as are accounts of journeys by fare-paying passengers, or on commercial craft.

Notes on conventions, abbreviations and locations will be found in the Introduction to Chapter 7.

ANDERSON, R[oger] C[harles], b. 1883.
Canoeing and camping adventures: being an account of three cruises in northern waters.
London: Gilbert-Wood, 1910. xvi (incl. photo frontis.), 192; illd., maps.   B
     Two cruises on inland waterways of Sweden by sailing canoe and dinghy.
BADEN-POWELL, Warington, 1847–1921.
Canoe travelling: log of a cruise on the Baltic, and practical hints on building and fitting canoes.
London: Smith *et al.*, 1871. Frontis., xiv, 172, [4] (pub. cat.), [1] pl., fldg. map; illd.   B
     Includes canal, river and lake travelling in Sweden.
BENNETT, [Enoch] Arnold, 1867–1931.
From the log of the Velsa.
New York: Century, 1914. Frontis., [viii], 307; illd.   N, P
   U.K. edn. by Chatto & Windus (London) 1920. Facs. of 1st edn. by Books for Libraries Press
   (New York) 1975.
     Includes motor yacht trips through Belgium and Netherlands.
BIGELOW, Poultney, 1855–1954.
Paddles and politics down the Danube.
New York: Webster, 1892. Frontis., 253; illd.   B
   U.K. edn. by Cassell (London) 1892.
     Same trip as described by MILLETT (q.v.).
BIRD, A[rthur] F[rederick] Ryder.
Boating in Bavaria, Austria, and Bohemia, down the Danube, Moldau, and Elbe.
Hull: Andrews, 1893. [iii], frontis., 160; illd., maps.   B
BROUGHAM, Reginald [Thomas Dudley].
A cruise on Friesland 'Broads'.
London: Ward & Downey, 1891. xii, 207, fldg. map; illd.   B
   Also 'new edn.' 1891.
CHASE, Mrs. Lewis.
A vagabond voyage through Brittany.
London: Hutchinson, 1915. Frontis., vii, 316, [63] pl.; map endpprs.   B
     By dinghy.
CHATER, Melville.
Across the Midi in a canoe: two Americans paddle along the canals of southern France from the Atlantic to the Mediterranean.
National Geographic Mag., 52 (2), August 1927, 127–167; illd., map.   B

CHATER, Melville.
The Danube, highway of races.
National Geographic Mag., **56** (6), December 1929, 643–697; illd.   B
    From Turnu Severin to Regensburg by launch and tug.
CHATER, Melville.
Through the back doors of Belgium: artist and author paddle for three weeks along 200 miles of Low-Countries canals in a Canadian canoe.
National Geographic Mag., **47** (5), May 1925, 499–540; illd., map.   B
CHATER, Melville.
Through the back doors of France: a seven weeks' voyage in a Canadian canoe from St Malo, through Brittany and the château country, to Paris.
National Geographic Mag., **44** (1), July 1923, 1–51; illd., map.   B
CHATER, Melville.
Two canoe gypsies: their eight-hundred-mile canal voyage through Belgium, Brittany, Touraine, Gascony and Languedoc: being an account of backdoors life on bargeman's highway.
New York: Brewer *et al.*, 1932. Frontis. ix, 230, [48] pl., [3] maps.   B
    U.K. edn. by Lane (London) 1933, under shortened title.
CHATTERTON, E[dward] Keble, 1878–1944.
Through Brittany in *Charmina*: from Torbay to the Bay of Biscay in a 6-tonner.
London: Rich & Cowan, 1933. Frontis., xiii, 256, [8] pl.; illd., map endpprs.   B
    3rd imp. 1933.
CHATTERTON, E[dward] Keble, 1878–1944.
Through Holland in the *Vivette*: the cruise of a 4-tonner from the Solent to the Zuyder Zee, through the Dutch waterways.
London: Seeley, Service, 1913. Frontis., 248, [8] (pub. cat.), [15] pl.; illd., maps, map endpprs.   B
    U.S. edn. by Lippincott (Philadelphia) 1913.
CHATTERTON, E[dward] Keble, 1878–1944.
To the Mediterranean in *Charmina*.
London: Rich & Cowan, 1934. Frontis., xii, 239, [1] (pub. cat.), [16] pl.; map.   B
    Nantes to Toulon via the Gironde and canals.
COLES, K[aines] Adlard.
In broken water: being the adventures of a six-tonner through Holland and among the Frisian and Danish islands.
London: Seeley *et al.*, 1925. Frontis., 183, fldg. chart, [12] (ads.); illd., maps.   B
    2nd edn. by Coles (London) 1967.
        Includes Dutch inland waterways and Kiel Canal.
DAVIES, G[eorge] Christopher, 1849–1922.
Cruising in the Netherlands. A handbook to certain of the rivers and canals of Holland, Friesland, and the north of Belgium.
London: Jarrold, [1894]. 208, fldg. map; illd.   B
    More of a log than a handbook.
DAVIES, G[eorge] Christopher, 1849–1922.
On Dutch waterways. The cruise of the S.S. *Atalanta* on the rivers & canals of Holland & the north of Belgium.
London: Jarrold, [1887]. x, 379, [1] (pub. cat.); illd.   B
    Apparently 2 issues: one with frontis., leaf ix/x listing illns., 11 (paginated) plates; the other has frontis., no leaf ix/x, and no plates; but has 11 gaps in pagination.
DONNER, Mrs. Jos[ef] Alex[ander].
Down the Danube in an open boat.
London: Blackwood, [*c.* 1895]. Frontis., 145, [14] (pub. cat.), [7] pl.   B
DOUGHTY, Henry Montagu.
Friesland meres and through the Netherlands: the voyage of a family in a Norfolk wherry.
London: Low *et al.*, 1889. Frontis, xii, 359, [5] pl., [2] fldg. maps; illd.   B
    3rd edn. 1890. 4th edn. by Jarrold (London) 1900.

DOUGHTY, H[enry] M[ontagu].
Our wherry in Wendish lands: from Friesland, through the Mecklenburg lakes, to Bohemia.
London: Jarrold, [c. 1892]. 406, [2] (pub. cat.), [4] fldg. maps; illd.   P, N
    2 pairs of leaves numbered 13–16.
    2nd edn. 1893.
[FARNELL, Lewis R., 1856–1934].
An Englishman's adventures on German rivers aus dem Tagebuche von L. Farnell. Herausgegeben von
Dr. A. Hamann, M.A., Oberlehrer.
Berlin: Simion, 1891. 97.   P
    English text, edited for German readers.
      Rowing trips, including Danube to Vienna.
FARSON, [James Scott] Negley, 1890–1960.
Sailing across Europe.
New York: Century, 1926. Frontis., xv, 354, [62] pl. fldg. map.   P, N
    Another edn. (1926) has only [6] pl. no fldg. map. U.K. edn. by Hutchinson (London) [1926].
    Unilld. edns. in Cape's Travellers' Lib. (London) 1928, 1936.
      By motor yacht from North Sea to Black Sea via Rhine, Ludwig's Canal, Danube. Later
      described in his autobiography 'The way of a transgressor', Gollancz (London) 1935.
FORESTER, C[ecil] S[cott]. 1899–1966.
The *Annie Marble* in Germany.
London: Lane, 1930. Frontis, vii, 311, [8] (pub. cat.), [19] pl., [2] fldg. maps.   B
    By motorised dinghy as far east as river Oder.
FORESTER, C[ecil] S[cott], 1899–1966.
The voyage of the *Annie Marble*.
London: Lane, 1929. Frontis, xi, 277, [6] (pub. cat.), [20] pl; map.   B
    By motorised dinghy up Seine, through canals, down Loire.
GORDON, James H[enry] H[amilton], 1845–1868.
A canoe voyage in the *ΡΟΘΙΟΝ.*
Cambridge: Metcalfe, 1868. [iii], frontis., 74, [1] pl.   B
    Reprinted from *The Light Blue* for private circulation.
      Solo journey in France and Germany.
HAMERTON, Philip Gilbert, 1834–1894.
A canoe voyage: etchings . . . in six monthly parts, each part containing six etchings.
London: Colnaghi, 1867.   P
HAMERTON, Philip Gilbert, 1834–1894.
Exploration of the Arroux, a canoe voyage, etchings in four monthly parts, each part containing
six etchings.
London: Colnaghi, 1887.   N
HAMERTON, Philip Gilbert, 1834–1894.
The Saône: a summer voyage.
London: Seeley, 1887. xx, 368, [16] pl. [4] maps; illd.   B
    Also simultaneous large paper edn.
    U.S. edn. by Roberts (Boston). 1888. 2nd U.S. edn., 1897, under title 'A summer voyage on the
    River Saône'.
      By berrichon and sailing catamaran.
HAMERTON, Philip Gilbert, 1834–1894.
The unknown river.
London: Seeley *et al.*, 1871. vi, 58, 2 (pub. cat.), [36] pl.   B
    This is 28 cm. high; in 1872, Roberts (Boston, U.S.A.) published a 23 cm. edn. A 19 cm. edn.
    by Seeley *et al.* in 1874 had only [8] plates.
      By paper canoe down Arroux, tributary of Loire.
[HARVEY, Edmund George, 1828–1884.]
Our cruise in the *Undine*: the journal of an English pair-oar expedition through France, Baden,
Rhenish Bavaria, Prussia, and Belgium. By the Captain.
London: Parker, 1854. Frontis., vii, 156, 4 (pub. cat.), [9] pl., fldg. map.   B

HAVARD, Henry, 1838–1921.
The dead cities of the Zuyder Zee: a voyage to the picturesque side of Holland [translated] from the French . . . by Annie Wood.
London: Bentley, 1875. Frontis., xiii, 363, [9] pl.   B
    New (revd.) edn. 1876.
        By sailing boat.
HAVARD, Henry, 1838–1921.
The heart of Holland. Translated by Mrs. Cashel Hoey.
London: Bentley, 1880. Frontis., ix, 386, [7] pl.   B
    U.S. edn. by Harper (New York) 1880.
        By sailing boat.
HEAVISIDE, George, and BENNETT, J. Edwin.
Canoe cruise in central & northern Germany, on the Fulda, Schwalm, Werra, Weser, and Geeste, in 1874.
Leamington: Vincent, 1875. viii, 175.   B
HUGHES, Robert Edgar.
Two summer cruises with the Baltic Fleet, in 1854–5. Being the log of the *Pet* yacht, 8 tons, R.T.Y.C.
London: Smith *et al.*, 1855. Frontis., viii, 333, 16 (pub. cat.), [3] pl. [2] fldg. charts.   B
    2nd edn. 1856.
        13–23: transit of Kiel Canal.
IRVING, Laurence [Henry Forster], b. 1897.
Windmills and waterways: the log of a summer cruise through Holland.
London: Heinemann, 1927. Frontis., viii, 291; illd., map endpprs.   B
KNIGHT, E[dward] F[rederick], 1852–1925.
The *Falcon* on the Baltic. A coasting voyage from Hammersmith to Copenhagen in a three-ton yacht.
London: Allen, 1889. Frontis., [vi], 309, [2] (pub. cat.), [9] pl. fldg. map.   B
    2nd edn. 1892. New edns. by Longmans (London) 1896, 1902. Reprinted in Mariners Lib., Hart-Davis (London) 1951.
        Inland through Netherlands and Flensburg.
KOEBEL, Arthur F.
Dinghey dawdle: Danubian and other. From Thames to Black Sea.
London: Cox (printer), 1902. vii, 79, [5] maps (1 fldg.).   B
        Via Rhine by dinghy and yawl.
[LEES, James Arthur, and CLUTTERBUCK, Walter J.]
Three in Norway. By two of them.
London: Longmans *et al.*, 1882. xv, 341, [8] pl. fldg. map; illd.   B
    4th edn. 1888. U.S.A. edn. by Porter & Coates (Philadelphia) 1883. Edns. in English by Heinemann & Balestier (Leipzig) 1892, and Brakhaus (Leipzig) 1925. Norwegian edn. by Tanum (Oslo) 1949.
        Canoeing.
LLOYD, M[ontague] A.
Up the Seine and down the Rhine: being an account of a 2000-mile cruise in a 21-ft. motor cruiser.
London: Imray *et al.*, 1938. [viii], 102, illd., map.   B
    2nd (unilld.) edn. 1950.
        Includes Belgian waterways.
The log of the *Water Lily*.
See MANSFIELD, Robert Blachford.
MACDONELL, Arthur A[nthony], 1854–1930.
Camping voyages on German rivers.
London: Stanford, 1890. Frontis., xvi, 278, [2] (pub. cat.), [20] maps.   B
MACGREGOR, J[ohn], 1825–1892.
The *Rob Roy* on the Baltic: a canoe cruise through Norway, Sweden, Denmark,Sleswig, Holstein, the North Sea, and the Baltic.
London: Low *et al.*, 1867. Frontis., [i], viii, 312, 16 (pub. cat.), 4 maps; illd.   B
    10th edn. 1894. U.S. edn. by Roberts (Boston) [1872].

MACGREGOR, J[ohn] , 1825–1902.
A thousand miles in the *Rob Roy* canoe on rivers and lakes of Europe.
London: Low *et al.*, 1866. viii, 318, 16 (pub. cat.), fldg. map; illd.   N
   1st edn. not seen.
   19th edn. 1892. At least 21 edns. U.S. edns. by Roberts (Boston) 1871, 1880.
   Facs. of 13th edn. (1881) by British Canoe Union (London) 1963.
MACGREGOR, John, 1825–1892.
The voyage alone in the yawl *Rob Roy,* from London to Paris, and back by Havre, the Isle of Wight,
South Coast, &c.
London: Low *et al.*, 1867. Frontis., xii, 335, 24 (pub. cat.); illd., maps.   B
   U.S. edn. by Roberts (Boston) 1880.
   Reprinted in Mariners Lib., Hart-Davis (London) 1954.
[MANSFIELD, Robert Blachford, 1824–1908.]
The log of the *Water Lily* (four-oared Thames gig) during a rowing excursion on the Rhine, and other
streams of Germany. By an Oxford Man and a Wykehamist.
London: Parker, 1852. Map, iv, 50.   B
    The original intention, not achieved, was to publish this in a magazine.
MANSFIELD, Robert Blachford, 1824–1908.
The log of the *Water Lily* (Thames gig), during two cruises, in the summers of 1851–2, on the Rhine,
Neckar, Main, Moselle, Danube, and other streams of Germany.
London: Cooke, 1854 (2nd edn.). [ii] , iv, 124; illd., map.   B
   The 'first edn.' took the form of two separate books, published anonymously: 'The log of the
   Water Lily' (q.v.) and 'The *Water Lily* on the Danube' (q.v.). 3rd edn. by Ingram (London) 1854.
   [4th?] edn. by Tauchnitz (Leipzig) 1854.
MANSFIELD, Robert Blachford, 1824–1908.
The log of the *Water Lily* during three cruises on the Rhine, Neckar, Main, Moselle, Danube, Saône,
and Rhone.
London: Hotten, [1873] (5th edn.); xii, [7] — 182, [1] (pub. cat.); illd., map.   P, N
   See previous entry for earlier edns.
   [6th?] edn. by Routledge (London) [1877].
[MANSFIELD, Robert Blachford, 1824–1908.]
The *Water Lily* on the Danube: being a brief account of the perils of a pair-oar during a voyage from
Lambeth to Pesth. By the author of the 'Log of the *Water Lily*' . . .
London: Parker, 1853. Pictorial title, xi, map, 216, 4 (pub. cat.), [7] pl.   B
   Included in all subsequent edns. of 'The log of the *Water Lily*' (*see* above).
MAXWELL, Donald, 1877–1936.
A cruise across Europe: notes on a freshwater voyage from Holland to the Black Sea.
London: Lane, 1907. 255, [4] (pub. cat.); illd.   B
   2nd edn. 1925.
MAXWELL, Donald, 1877–1936.
The log of the *Griffin*: the story of a cruise from the Alps to the Thames.
London; Lane, 1905. 305, [2] (pub. cat.); illd.   B
MILLET, F[rancis] D[avis] , 1846–1912.
The Danube: from the Black Forest to the Black Sea.
London: Osgood, McIlvaine, 1892. Frontis., xv, 329; illd., maps   B
   U.S. edn. by Harper (New York) 1893.
   By canoe. Same trip as described by BIGELOW (q.v.).
MINSHALL, Merlin.
By sail across Europe.
National Geographic Mag., 71 (5), May 1937, 533–567; illd., maps.   B
   Le Havre – Paris – Strasbourg – Ludwig's Canal – Danube, by motorised Dutch sailing barge.
   Later described in his autobiography 'Guilt-edged', Bachman & Turner (London) 1975.
MOENS, W[illiam] J[ohn] C[harles] , 1833–1904.
Through France and Belgium, by river and canal in the steam yacht *Ytene*.
London: Hurst & Blackett, 1876. Frontis., xii, 310, 16 (pub. cat.).   B

MOLLOY, J[ames] L[yman] , 1837–1909.
Our autumn holiday on French rivers.
London: Bradbury, Agnew, 1874. xviii, 391; illd.   B
   2nd edn., nd, smaller format.
     U.S. edns. by Roberts (Boston) 1879, 1885.
       Rowing on Seine and Loire.

Our cruise in the *Undine*.
See HARVEY, Edmund George.

PEARS, Charles, b. 1873.
From the Thames to the Netherlands: a voyage in the waterways of Zealand & down the Belgian
coast.
London: Chatto & Windus, 1914. Frontis., xvi, 211, [31] pl., [3] maps.   B
     By yacht.

PEARS, Charles, b. 1873.
Going foreign: a volume of information for those about to cruise abroad in small yachts, with speci-
men voyages in Holland, Belgium and France.
London: Arnold, 1933. 224; illd., maps.   B

POWELL, James.
Our boating trip from Bordeaux to Paris.
London: *The Field*, 1880. [vii] , 40, [4] (ads.). /
   Reprinted from *The Field*.
     By skiff.

RAVEN-HART, R[owland James Milleville] .
Canoe errant.
London: Murray, 1935. Frontis., xv, 291, [15] pl.; map endpprs.   B
     Extensive trips in Europe.

RAVEN-HART, R[owland James Milleville] .
Down the Rhône.
Geographical Mag., **IX** (3), July 1939, 177–190; illd., map.   B
     By canoe.

RICHARDSON, Leslie, b. 1885.
Brittany and the Loire.
London: Bles, 1927. Frontis., 288, [11] pl.; map endppr.   B
     Inland and coastal cruising.

RICHARDSON, Leslie, b. 1885.
Motor cruising in France from Brittany to the Riviera.
London: Bles, 1926. Frontis., 286, [11] pl.; map endpprs.   B
   2nd edn. 1932. U.S. edn. by Mifflin (Boston) 1926.
     Inland from Bordeaux to Mediterranean.

RISING, T[homas Craske] , and RISING, T[ean] .
Kingfisher abroad.
London: Cape, 1938. Frontis., 247, [15] pl.; maps.   B
     Canoe trips in Germany, and down Danube to Budapest.

ROBINSON, Charles E[dmund Newton] , 1853–1913.
The cruise of the *Widgeon*. 700 miles in a ten-ton yawl. From Swanage to Hamburg, through the
Dutch canals and the Zuyder Zee, German Ocean, and River Elbe.
London: Chapman & Hall, 1876. Frontis., xv, 268, [4] (pub. cat.), [3] pl.   B
   2nd edn. 1877.

ROWLAND, Henry C[ottrell] , 1874–1933.
Across Europe in a motor boat: a chronicle of the adventures of the motor boat *Beaver* on a voyage
of nearly seven thousand miles through Europe by way of the Seine, the Rhine, The Danube, and
the Black Sea.
New York: Appleton, 1908. Frontis., xi, 304; drgs.   B
   U.K. edn. by Appleton (London & New York) 1915.

RUMBOLD, C[harles] E[dmund] A[rden] L[aw], b. 1872.
Yacht cruising on inland waterways, from the Baltic to the Mediterranean.
London: Oxford University Press, 1935. Frontis., xiii, 200, XI pl., fldg. pl.; illd., maps.   B
   Supplement issued November 1936: 2 p., card covers.

Six weeks on the Loire.
See STRUTT, Elizabeth.

SPEED, H[arry] Fiennes, d. 1925.
Cruises in small yachts and big canoes; or notes from the log of the *Watersnake*, in Holland and on
the south coast . . .
London: Norie & Wilson, 1883. Frontis. chart, viii, 288, [9] charts (2 fldg.); illd.   B
   See next entry for 2nd edn.
     1–73 describe cruise including waterways of Holland and Belgium.

SPEED, Harry Fiennes, and SPEED, Maude.
Cruises in small yachts, by Harry Fiennes Speed: and a continuation, entitled More Cruises, by
Maude Speed.
London: Imray *et al.*, 1926 (2nd edn.). Frontis., [xiv], 355, [14] pl. [4] charts (1 fldg.); illd.   B
   See previous entry for 1st edn.
     'More Cruises' includes an inland cruise in Netherlands.

STEVENSON, Robert Louis.
An inland voyage.
London: Kegan Paul, 1878. Frontis., x, 237.   B
   Numerous editions.
     Canoeing in Belgium and France.

[STRUTT, Elizabeth.]
Six weeks on the Loire . . .
London: Simpkin & Marshall, 1833. Frontis., vii, 408, [3] pl.   B
     69–272: cruise from Tours to Nantes.

SUFFLING, Ernest R[ichard].
A cruise on the Friesland meres.
London: Gill, 1894. 50, 16 (ads.); illd.   B

THORPE, [Thomas] Edward, 1845–1925.
The Seine from Havre to Paris.
London: Macmillan, 1913. Frontis., xxi, 493, 15+[2] maps, [2] (pub. cat.); illd.   B
   By steam yacht.

Three in Norway. By two of them.
See LEES and CLUTTERBUCK.

TOMALIN, H. F.
Three vagabonds in Friesland with a yacht & a camera.
London: Simpkin *et al.*, 1907. Frontis., [xiv], 251, xxvi (App.); illd., map endpprs.   B
   Also a limited edn. of 175 by Adams Bros. (London) 1907. 2nd edn. 1907.
     U.S. edn. by Dutton (New York) 1907.

TOMLINSON, Harry, and POWELL, James.
Camp life on the Weser.
London: *The Country*, 1879. viii, 53, [2] (ads.); illd.   B
     By skiff from Bremen to Minden and back.

TUDSBERY, M[armaduke] T[udsbery].
The sea to the Schwartzwald and back.
London: for private circulation, 1934. Frontis., [ii], 44, [3] pl.   B
   2nd edn., for private circulation, 1940.
     Rhine and Neckar by motor launch.

VAN TIL, William [Andrew].
The Danube flows through Fascism: nine hundred miles in a fold-boat.
New York: Scribner, 1938. Frontis., xiii, 301, [30] pl.; maps.   B
     Down to Belgrade.

WARING, George E[dwin], 1833–1898.
The bride of the Rhine. Two hundred miles in a Mosel row-boat.
Boston, Mass.: Osgood, 1878. [i] (pub. cat.), frontis., 312; illd., maps, fldg. leaf 193/6.   B

WARREN, E[dward] Prioleau, and CLEVERLY, C. F. M.
The wanderings of the *Beetle*.
London: Griffith *et al.*, 1885. vi, 178; illd.   B
    Title-page gives date 1835.
       Rowing in Belgium and France.

The *Water Lily* on the Danube.
See MANSFIELD, Robert Blachford.

CHAPTER 5

WHAT IS CONSERVATION?

'THAT'S WHERE WORSEY'S used to trade' and 'the canal used to go under that bridge and into Harris's Yard — built a good boat they did!' The speaker was Malcolm Braine, and I was journeying a few years ago through the Birmingham Canal Navigations (BCN) and the Black Country in his company discovering, largely by detective work and his memory, various, almost vanished, Black Country side-slip yards where joey dayboats and other craft were traditionally constructed and repaired.

My task was to help design and recreate as a live exhibit an authentic boat-building and repair yard as it would have been at the turn of the century as part of the

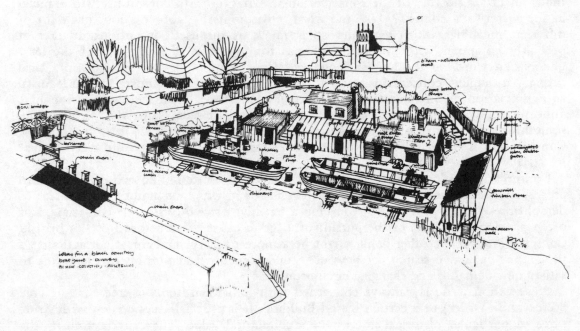

Figure 5.1. Ideas for a re-created Black Country boatyard for the Black Country Museum at Dudley.

interpretative displays of the Black Country Museum in Dudley. It was an exercise in the kind of historical preservation/interpretation that I consider to be immensely interesting. Malcolm explained that rather than 'burning their boats' (once they stopped floating), expedience had demanded that BCN joeyboats be turned on their sides to become 'boat bottom' buildings, and so a form of structure was actually created out of thrift and economy — or perhaps downright meanness! These sheds were clear evidence of good old BCN attitudes, and it was something I had seen many times before on quite different canals. How on the Oxford Canal a certain half-lapped nailing of slates enabled far fewer to be used as a roof covering, how 'rat-trap' bonding also *saved* bricks because they would be laid not on a bed face but on edge, how brick wall copings often had a 'frog' created in their undersides (actually formed by a hand scrape whilst the clay was still wet) in order to *save* clay and facilititate laying. So whether born out of 'good housekeeping', or any other cost-saving motive, these fascinating features had made a real and actual contribution to the architectural built form of our waterways; they demonstrated, too, that Conservation (in this case of scarce resources) isn't just a new fad or a thing suddenly discovered in the 1970s.

But this was not the only thing that is frequently explained to the observer of the waterway scene. Interesting and fascinating lessons abound, and inspiration is never far away from those who can go beyond just 'looking', in order to see and understand. I remember once listening to a spokesman for the Society for the Preservation of Ancient Buildings playfully deride the 'do-gooder' Conservationist (whom he described as 'the strawberry jam merchant') by effectively extolling the virtues of no-nonsense, outright, honest to goodness *Preservation*. In so doing he cleverly exposed the sometimes uncertain posture of the conservationist. What on earth does Conservation mean in the waterways context? Like the word 'environment', 'conservation' has a lot of different meanings which vary almost as much as the user's intentions. In fact, it probably competes with the word 'marina' for the prize of being one of the least understood words in the English language. 'Conservation', to me, is about practical caring – a tangible concern for waterside buildings, places and landscape that leads to an appreciation of, and an attraction for, their appearance, character, their personality, atmosphere, scale and form. The word to some might seem somewhat fashionable, something to do with tarting things up – facelifting, or a 'lipstick job' – but because Conservation is often misunderstood I think it might help if we could discuss what it ISN'T.

Conservation isn't the same as Preservation (and so quite unlike our BCN boatyard preservation exercise). Preservation, or the call for it, often emanates from a threat of demolition or from pressure to redevelop a sensitive piece of waterside landscape. It is often a defensive battle, or the pursuit of a 'good cause' – so beloved of the British. Conservation, on the other hand, might be a more positive stance, not a reaction to a threat but an active search for new uses, or an exciting adaptation of old forms to satisfy new demands and changing circumstances.

Preservation is nearly always concerned with individual items: a tree (with a Tree Preservation Order) or a certain Listed Building. It is especially concerned with structures of particular historical interest or architectural value. The criteria used by the Department of the Environment (DOE) in their Statutory List always relate to one or both of these value ratings, and if alteration or demolition of a building is envisaged then it understandably becomes the subject of official control. Important obligations then follow for the owner to meet. Conservation, however, is concerned with overall 'spaces and places', with what might sometimes be called the 'group value' of buildings together with their surroundings and what goes on in them: a *totality* that, by consent, represents something important needing to be looked after. It is not about museum pieces, empty shells, nor architectural gems in a sea of mediocrity, but about assets that can be converted and re-used for practical and economically sound reasons. A conservationist, then, is not only interested in the fabric itself, but also in the people who 'spun or wove it' and their successors, who have to live with it and use it both today and tomorrow.

Perhaps like fine art, Preservation requires, by definition, an articulate and an expert judgement of merit, whereas Conservation is about ordinary sensations and feelings that are more down to earth and understandable. Spaces say things in a less complicated way than the, often, more academic message of buildings — and the words invariably seem friendly.

For our waterway environs, an important role is instantly recognised. Canals and their environs act importantly and psychologically in our lives by providing a friendly reassurance in a world of change. They offer familiar landmarks. Therefore, before we knock down or destroy, we must realise that we uproot the origins of our waterways' heritage at our cost!

Preservation often says 'No' to redevelopment, because it insists on retention regardless of circumstances. Conservation might say 'Let's keep what we can whilst ensuring that new replacement buildings, waterway equipment or canalside landscape should be designed in harmony with a canal's particular architectural character'. It might be something about good manners or courtesy and it certainly needs care and judgement to get it right. Conservation then, in the sense of the word I wish to explore here, primarily concerns the built form of our waterways and their landscape context. What opportunities and dangers are there?

For anyone interested in the canals of this country, the bookshelf is now full (and bending under the weight) of some exciting tomes. They guide, stimulate and explain. Not so in my young day! Three authors were then the only essential reading — Eric de Maré, L. T. C. Rolt, and Charles Hadfield — an architect, an engineer and an historian respectively. From 30 years on, de Maré's images in his book *Canals of England* somehow seem to have acquired an important documentary significance, and certainly in his splendid black and white photographs he imparted an appropriately dramatic impact into my life, of a kind that would do justice to great theatre.

Some of his camera work triggers those half-forgotten images of boyhood, such as when in the 1950s, in home-knitted train-spotter's pullover, I stood on the lattice footbridge at Hatton Junction glimpsing through the steam, without realising it, the 'closing overs' of narrowboat carrying on the Grand Union Canal. De Maré's work demonstrated too that the fabric of our waterways was not just about engineering structures or canal buildings on their own, but that the subtlety, the harmony, the often delightful landscape context, were *all* part of what could perhaps be more aptly described as a 'historic landscape'. His commentary about places, and his remarks about the people who inhabited this unknown stage, meant, for me, that the gate to the secret garden had been left open. With his tremendously evocative images de Maré had helped me to stumble upon it all.

In *Canals of England*, here was an architect like myself, not just talking about a boat journey, but also able to explain that *design* wasn't only about the super-ficial appearance of historical structures. With his beautiful pictures he illustrated the simple origins and the meanings of the Functional Tradition, how commercial needs and the expediences of the time had all helped to create the form, the scale and appearance of the waterway scene. Today, the baton has been taken up by Derek Pratt in his photographic work with Anthony Burton.

I began to understand how use and wear had imparted their own delightful patina to be added to that of erosion, the weather and the passage of time to create fascinating effects. I was introduced also to a friendly unassuming world, an environment that was charmingly straightforward, sometimes simple, often somewhat neglected and slightly rough at the edges, a world that you took as you found it, a world that was frequently and visually arresting, but a scene that asked questions! De Maré, and Tom Rolt too, suggested, by implication, just as with anything of value, that there were (and are!) real dangers that could quite easily devastate and devalue. How easy has it become for us to ruin the magic appeal that we remember was possessed by places we once knew; looking through their books now, I am dismayed and outraged by how much has been lost and neglected, how many opportunities missed or ignored. Why is this? Are we, the Great British Public, so preoccupied with communicating with dogs that we are not able to commune with buildings and views and places? (I am tempted to think that had each of the demolished buildings lost since de Maré and Rolt cruised the waterways of this country been a Dogs' Home, then they would have somehow remained intact!) British attitudes to taste, design and environmental awareness are frequently elusive. What is certain is that when old, familiar things are removed, we miss them! Removal often underlines their value — when it is too late. Look around you!

Value and Character

I suppose the most humiliating thing about canals and their environs is that most were entirely conceived and designed *without* an Architect in sight! But when thumbs did the ruling and instructions were in the pattern books, the disciplines of choice and the expediences of costs and dividends all conspired to create a subconscious regard for taste and good judgement. Architects weren't needed. Not so today, when we seem completely spoiled for choice and 'free' from such design and building constraints! The strength and permanence of it all was something too that I still find remarkable, in contrast to our present-day attitudes that are so conditioned by flexibility and change. Perhaps this is why so many of the features in the canal scene, which de Maré showed and Hadfield explained, still evoke such a splendid aura of confidence — a feeling that is strangely comforting in our own difficult times! It is quite sobering to think that, although early canals were created literally out of the landscape and that as a result they fit in as man-made structures delightfully well in the natural scene, they did enable, for the first time, extremely alien building materials to be imported from outside the immediately local district. I am sure it is things like this, together with an understanding of the enormous time span and the period in which canals were 'in building', the variety of contractors, and the changes of this or that Engineer, that somehow created delightful idiosyncracies and inconsistencies (and several architectural jokes) that never fail to puzzle, fascinate and enthral us. Our waterways have since become a very rich part of our industrial heritage, and perhaps the fact that they have survived as they have is, in itself, a phenomenon that we find impressive.

Well, if the patient has survived, what state of health is *she* in? (I am certainly convinced that Temple Thurston was right in his book *The 'Flower of Gloster'* when he spoke of the Water Deity in female terms! For close to our regard of boats belonging to the gentle sex must follow a similar compliment to our navigations. The female is certainly prominent on a gentle misty day in Shropshire, or in the autumn on the Oxford Canal, or in the springtime at Tixall. But what about a sharp February morning at Wigan Pier or a brisk November day at Smethwick? Yes! Female they are. Do our waterways exhibit problems and needs, a bit like those of an old maiden aunt who has fallen on hard times? There suddenly seems a lot of life left in her and much of this is explained in that millions of people increasingly *expect* to inherit something. Indeed, far from merely having expectations, many are *possessed* to the point of obsession. A great, dedicated, voluntary movement (so peculiarly and endemically British, because of our usual national indifference) has developed since the Second World War and thousands of people now work to restore derelict waterways to navigability. But what do we *do* with them once they have been restored?

If it is a matter of Conservation (practical caring) then what about the components that make up the picture, the features that perhaps we take for granted? As important as the immediate waterway engineering structures (the locks, the tunnels, the aqueducts, the reservoirs, and the channel itself) are the canal *environs*, the buildings, the yards, the wharves, the basins, the lock cottages, the public houses, the lock lobbies, and the toll houses – the very fabric of the industrial transport enterprise. What is happening to these buildings, this architectural backdrop, in both the urban and rural landscape context? The backdrop is crucial because it actually makes up the waterway environment which we all love, the thing we are all increasingly wishing to enjoy. How is this environment sustaining leisure pressures? How can it perform a role for which it was never designed? What is the appeal? Does everything need preserving? How appropriate is its Conservation?

Canal Contrasts in Town and Country

It is astonishing that in our towns and cities two 'worlds' can and do exist, side by side, one full of the bustle and the stress of the 20th century, and the other, the waterway environment, a complete antidote, a delightful, friendly unassuming escape route. The 'detachment' of canals in towns, illustrated so well in Braithwaite's book *Canals in Towns* (Black, 1976) explains their frequent surprise and delight, and the excitement of exploration and the discovery of familiar landmarks from a quite unrealised and unusual angle. This secret and somewhat introverted sense of enclosure is a feature of great importance, as is the relatively undisturbed corridor of the towing paths and their immediate environs of cuttings, embankments, wharfs, canal arms and basins, often possessing delightful landscape and ecological value. It is a magic scene.

In some ways this 'canalscape' can be said to be an historical accident of minimal maintenance. Trees and shrubs have been allowed to grow and to colonise waterside land, hedgerows have matured, and wild flowers cover the margin at the waterside and at the rear of the towing path. As a safe traffic-free environment it is enjoyed by fishermen, naturalists, boaters and school parties, and by walkers (perhaps exercising their dogs, now that many parks are dog-free!). Much of the enjoyment of these users is derived from the friendly unassuming quality of the towing path scene. It is not a leisure outlet that is elaborately manicured and cultivated. Users take it as they find it.

In areas deprived of open space, canals bring a much valued breath of the countryside into the lives of many. The canal corridor possesses an astonishing variety of experiences: surprise and delight, shock and encounter; it is a waterside landscape that often displays wilful neglect, and planning indifference, but occasionally encouraging evidence of care. Surroundings may vary but the consistent element running through the canal scene is the water itself, unifying everything around it, reflecting, shimmering, and gurgling, always magnetic, and attracting *use* as well as abuse!

It is hardly enough to treat the water as something just to sit by, and look at. Developments that include water simply as just a visual feature are, more often than not, totally disappointing; a shallow water channel is soon choked with lollipop sticks, fag ends and litter, and attempts to create such linear 'water parks' out of derelict canals have been visually and managerially disastrous! It certainly seems to be the case that for waterspace to be effective, it needs to be used and enjoyed fully so that many user groups can help provide effective management and maintenance by being 'unpaid caretakers'.

Problems and Opportunities

From Coventry Basin, in the 1960s, I cruised in the wake of Tom Rolt's *Cressy* in my own narrowboat home *Calstock* and *saw*. It didn't take long for me to realise that it is not just a question of money or the lack of it, but one of attitudes, *our* attitudes. We all see vandalism and graffiti in our canalside towns demonstrating the 'physical prowess' of the inhabitants or the particular skills of their football teams; we see neglect and indifference and attitudes that seem to dump the community's prejudices into the canal itself; we see appalling waterside developments that have totally ignored the exciting opportunities of an attractive waterside frontage; we see a canal network struggling to accommodate quite demanding leisure pressures; we see old buildings, many owned by the British Waterways Board (BWB), derelict and neglected; we see ugly new buildings built with a complete lack of courtesy and a total disregard to canal traditions. Does it have to be like this?

In 1967, I realised that something needed to be demonstrated, and so, through working in the City of Birmingham for three years, I was able to plan and design a waterside renewal scheme on behalf of the City Architect that began to illustrate some of the exciting possibilities – James Brindley Walk. This work was actually inspired by Gordon Cullen's tantalising drawing of a suggested urban canal treatment in *Canals of England* (p. 118) and the scheme does demonstrate a way of coming to terms with water in towns. The dreadful fencing around the basin was a result of hysteria whipped up to present the canal as a potential hazard *after* the scheme had been implemented. The biggest danger to young children is *not* an unprotected water's edge, but a lack of parental supervision, and this disappointing need for a toddler-proof barrier is an interesting illustration of how daring and unacceptably adventurous the scheme was felt to be at the time. Have our attitudes changed that much since? Do we still find it difficult to come to terms with water? Have we learnt to respond to it positively?

Following this experiment, I was engaged by Sir Frank Price (Chairman of BWB) in 1970 to develop these ideas as the Board's Architect/Planner, and in this capacity I have tried to foster and encourage a growing awareness within the Board of our environmental heritage, in a way that has influenced management decisions. It has been an eventful 14 years!

Changing Attitudes

We must change attitudes and conserve the appeal of the waterway environment. Whilst it is the Board's duty and policy to encourage recreation and amenity uses of their canals, how can we expect people to use and enjoy them fully if the canal environment becomes increasingly uninviting and often remains visually depressing? How can we expect tourists to visit the canals, if our historical heritage is consistently being visually eroded? Many canals in towns pass through areas where they are regarded as dangerous, as rubbish dumps or as targets for vandalism. However, attitudes are beginning to change and increasingly canals are used for school and college projects, and many young people are becoming aware of the significance of waterways in their environment. Many children, by looking at the industrial archaeology, the economic history, and the botanical and ecological value of the canals and their towing paths, are developing a new set of values and attitudes. Occasionally, perhaps, these supplant parental preconceptions that may have regarded canals as little more than a rubbish dump or 'somewhere to drown the dog when the licence becomes due'.

Attitudes towards safety may have also traditionally regarded the canals as an awkward boundary and something to be fenced off, but there is now a realisation that a safe waterway is one that is accessible, one that can be used and enjoyed fully. In the experience of young people canoeing and using small boats, much can be gained in their learning a respect for water, and care for their surroundings, and by their developing self-reliance and confidence.

So perhaps the future of our waterways lies in how we go about using our increasing leisure time, how young people can be motivated to use canals in a positive and exciting way, and perhaps how jobs and employment situations can be enhanced by measures that enable canals to help improve the 'work environment'. But whatever policies the British Waterways Board or Local Authorities may have, unless these can be adequately financed, they will never be sensibly realised, particularly in periods of public spending cuts. The development of leisure facilities and amenities on the canals will therefore, I think, need to centre to a large extent on how we can attract and foster private commercial investment and interest — indeed the very motivation that created our canals in the first place — perhaps in an exciting new form of partnership?

Environmental Enhancement

It is always helpful for somebody to set a modest example, and no better illustration of this can be seen than in the Board's own environmental enhancement and improvement schemes. Our waterways are certainly no place for large areas of tarmac and beautifully-mown grass, because that means the worst sort of 'municipal' sterility; in fact, the very thing from which most people are anxious to escape! A delicate balance, therefore, must be struck, so that normal repairs, environmental and operational improvements can appear consistent with the canal idiom, having regard to public safety of course, but also in a way in which they can be maintained with efficiency. Maintenance must be of a sufficiently high standard to avoid vandalism and abuse, and it is apparent from the Board's experience that when structures appear to be neglected they actually *attract* the attention of vandals!

In recent years, the BWB has attempted to build into its normal engineering maintenance programme a modest level of visual enhancement, and to complement these essential works wherever possible with small-scale amenity improvements. This is not merely a cosmetic exercise; painting requires to be done in any case in order to prevent rust, rot or decay. Several improvement schemes have largely been centred around lock flights, canal junctions and maintenance yards and although expenditure has necessarily been extremely modest, the visual impact at several specific locations can be seen to have had a successful effect. There is however much still to be done.

Early on, it became apparent that, to avoid inconsistency in design and planning issues, some sort of policy guidelines might be helpful. Many visually irritating matters had frequently occurred not because someone was deliberately indifferent but because small isolated decisions had never been looked at in an overall context. Thus emerged the *Waterway Environment Handbook* (BWB, 1972) which was accepted and endorsed by the Board as a policy and a means of rescuing the canal scene from ill-conceived design and the worst excesses of over-enthusiastic maintenance. The *Handbook* explained that often it was not so much what was done but what was *not* done that was important. A Civic Trust Award (1972) on the Regent's Canal endorsed this approach by applauding the Board's efforts 'for resolutely refusing to overstifle the canal vernacular'. Conserving our past therefore does not mean spending vast sums — because here we received an Award for doing virtually nothing! Thus a sensitive campaign designed to conserve and facelift the canal fabric was launched, and since 1972 its impact had been quite extensive. The fact that the Board has been able to do anything at all in this field is, I suppose, reasonably impressive when the various financial and legal constraints of the 1968 Transport Act are understood.

The foreword to the *Waterway Environment Handbook* says,

> As many of our waterways begin to serve a new industry — that of leisure — our attention is increasingly drawn, whatever our involvement, to the conservation of this fascinating appeal. Whether concerned with the normal maintenance of waterside structures, or the design and deployment of new amenity features, a whole range of attitudes towards design, colour and scale, can, within a series of constraints, allow the unique appeal of our waterways to flourish. By building in a consistency and order we can jointly prevent, perhaps, our waterways being devalued.
>
> This is what the *Handbook* is about — in many cases all that is needed is a coat of paint, and data sheets describe colours and locations — often so traditional as to be obvious to Waterways Staff and enthusiasts, but of value to other parties who share an interest and involvement in the waterway environment. Often, not only will an attractive result emerge, but also in presenting an efficient, well cared for undertaking, many of the pressures of vandalism and abuse will be avoided. The loose leaf information sheets describe some basic attitudes; local needs and conditions may require a particular approach that cannot possibly be included specifically in such a document . . . Solutions are suggested and recommended . . .
>
> Although originally produced in 1972 for the Board's own purposes, the *Waterway Environment Handbook* suggests answers to a wide range of environmental issues, often involving other parties. Because of this 'liaison' value, it has been widely circulated to most waterside Local Authorities, Civic Societies, Canal/River Societies, the boating industry, to Universities, and to architectural consultants and planners.

FIXED BRIDGES 01A

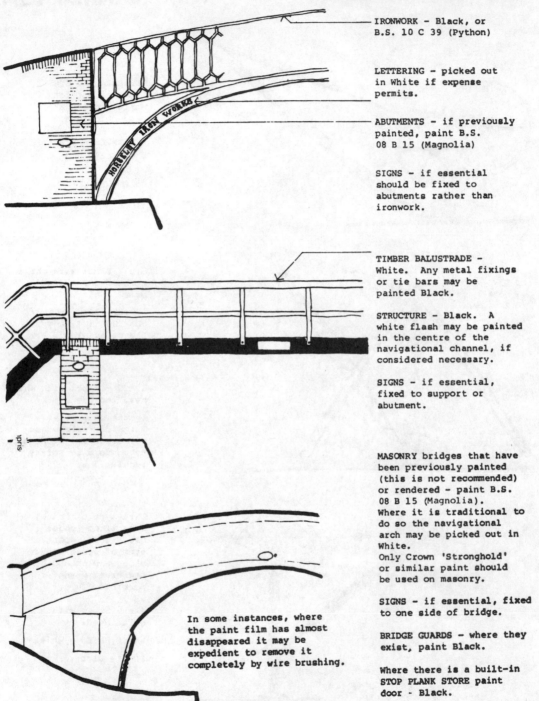

IRONWORK - Black, or
B.S. 10 C 39 (Python)

LETTERING - picked out
in White if expense
permits.

ABUTMENTS - if previously
painted, paint B.S.
08 B 15 (Magnolia)

SIGNS - if essential
should be fixed to
abutments rather than
ironwork.

TIMBER BALUSTRADE -
White. Any metal fixings
or tie bars may be
painted Black.

STRUCTURE - Black. A
white flash may be painted
in the centre of the
navigational channel, if
considered necessary.

SIGNS - if essential,
fixed to support or
abutment.

MASONRY bridges that have
been previously painted
(this is not recommended)
or rendered - paint B.S.
08 B 15 (Magnolia).
Where it is traditional to
do so the navigational
arch may be picked out in
White.
Only Crown 'Stronghold'
or similar paint should
be used on masonry.

SIGNS - if essential, fixed
to one side of bridge.

BRIDGE GUARDS - where they
exist, paint Black.

Where there is a built-in
STOP PLANK STORE paint
door - Black.

In some instances, where
the paint film has almost
disappeared it may be
expedient to remove it
completely by wire brushing.

Figure 5.2   'Fixed Bridges' — an extract from BWB's *Waterway Environment Handbook.*

YARD CRANES 09A

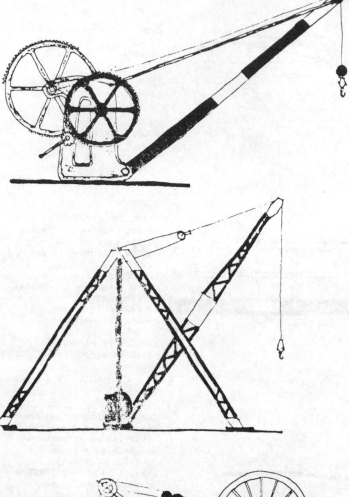

TIE BARS AND JIB –
Black except for
White section at
head and mid-span.

REMAINDER - Black

JIB - Black with white
at head and mid-span,
or all White.

CENTRE POSTS, MOTOR/
GEAR HOUSING, any
CAB paint Black.

STAYS - Black with White
head and foot.

In the case of very
large derricks other
colouring may be required
to conform to safety
regulations.

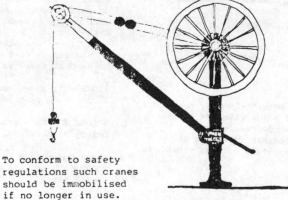

This type of crane,
even if no longer in
use, is an important
item of canal archae-
ology and should be
preserved whenever
possible.

JIB - Black with
White head.

LARGE WHEEL - White

POST - Black.

To conform to safety
regulations such cranes
should be immobilised
if no longer in use.

Figure 5.3. 'Yard Cranes' — an extract from BWB's *Waterway Environment Handbook*.

Resources will always be a vital factor but if a well chosen simple design is adopted then it is likely to be in complete harmony with the straightforward characteristics of the waterway scene.

A loose-leaf design manual approach seemed to be a good low-cost way of effectively disseminating information in a number of design situations that had several recurring common denominators. In 1972 (as now) what was needed was an understanding of a vocabulary of solutions, many inspired by tradition or precedent, which it would be appropriate to encourage and stimulate. Thus, out of necessity was born a manual that by 1975 had begun to transform the appearance of the waterways scene, and even led to the Board being presented with a special medal for a *continuing* contribution to conservation' in European Architectural Heritage Year.

Perhaps in an odd way the *Waterway Environment Handbook* with its advisory stance has also fostered several false assumptions and engendered a confidence towards Conservation and environmental awareness that is either non-existent or neglected! Maybe it demonstrates that it is no use having a lot of good ideas, if there does not exist a really effective means of implementing them. The problem with advice is that you can take or leave it! Look around you?

Unlike the Regional Water Authorities created in 1974, the Board has *no* duty at all placed upon it to conserve and enhance its buildings or to look after, and manage properly, the waterside landscape, and *certainly* it has no money for such a policy! Therefore, everything that the Board, as owner, has achieved in this field has been despite the financial difficulties and legal limitations. This has made it all the more necessary to influence, stimulate and encourage *other* parties to help the Board to achieve things which it is quite unable to do on its own. Another quite different answer would be to extend the Board's role and remit, to expand its environmental terms of reference, to establish agreed obligations, and then provide for the custodian the means to achieve such objectives. To 'expect', and then to deny the means, will always result in failure and missed opportunities. That something needs to be done is undeniable — but what?

Architectural Conservation

As I have tried to explain earlier, the 'group value' of buildings and structures, which individually might not be at all remarkable, may often be extremely important, requiring individual statutory protection by Listing or, in the case of places, by some sort of Conservation Area designation. The retention and conservation of the canal heritage must be an important priority because in this way tourism, commercial investment and helpful grant aid can be attracted. Public interest can be encouraged as well as the provision of facilities that might well assist with trade and employment in the local communities through whose area the waterways pass. It is always easy to be wise after the event, but unless we take note of previous mistakes, the continued loss of interesting or beautiful canalside buildings is even more serious. At sites where, regrettably, such demolition has occurred, it is especially important to replace lost buildings with sensitively conceived new buildings that complement the atmosphere that once was there.

A positive look at the conversion of old buildings into new uses would seem to be an essential part of retaining the fabric of the canal system, and in overall environment enhancement schemes the restoration of key existing waterside buildings and structures can often be an essential focus and feature. Unless, therefore, the British Waterways Board, Local Government and other agencies can, together, restore these derelict buildings and convert them to new uses, and unless they can together create historical trails and develop and take advantage of the industrial archaeological interest of our waterways, then we will surely lose the essential components that make our waterways so appealing. It is this appeal that *must* be conserved and enhanced, and it is an appeal that can earn money, too.

Landscape Conservation

Landscape conservation, management and the introduction of new planting has, of course, many tangible benefits; trees and shrubs can provide a living screen to unsightly industrial premises, delineate property, keep out stock, provide shelter and shade in exposed areas, create a habitat for wild life, and complement land reclamation schemes. It is essential to strengthen our commitment to canalside landscape conservation and management and to consider enhancement not as a hindrance to development, a cosmetic exercise, or an unnecessary maintenance operational burden, but as an important and integral part of the canal environment. Some forms of recreational activity, insensitive maintenance and pollution can reduce or obliterate the sensitive wild life value, while *careful* management and sensitive planning, together with improvements to water quality, can certainly enhance natural habitats. Perhaps it is necessary, therefore, for the Board to liaise and co-operate much more with Local Authorities, amenity groups, County Trusts for Nature Conservation, and Water Authorities in managing and maintaining the waterway environs appropriately so that this rather delicate landscape appeal is not destroyed. A 'rescue' team would not be necessary if damage was not created in the first place.

Listing and Conservation

Listing? It is perfectly clear that the statutory procedures leading to the Listing of canal structures that are recognised as being of architectural and/or historical interest have, in fact, created some rather bizarre inconsistencies throughout the canal network. Occasionally, quite different criteria have been used in order to evaluate the potential merit of respective structures and quite ludicrous situations have emerged on canals passing through different County and District Authority areas. A bollard or mooring post can be a Listed Structure in one particular District area, whereas in an adjoining area on the same canal, unique structures of immense historical importance are entirely omitted from the DOE List! Is the system in disrepute?

In these circumstances it might be far better for the British Waterways Board to be trusted to establish its own evaluation of important structures so that a list of priorities could be established and agreed with the DOE that would enable management and maintenance decisions decisions to be handled appropriately. There is always some alarm and concern about the implications of Listing in as much that BWB engineers take the view that a potentially greater degree of maintenance and upkeep is a distinct financial liability in the context of extremely limited financial resources. However, here surely is an excellent opportunity for the Board to demonstrate clearly the evidence of increased 'expectations' from local communities and argue that their own terms of reference have now been enlarged, implicitly, quite beyond the rather inadequate statutory duties of the 1968 Act. Clearly if the British public *expects* a policy of conservation and enhancement from a Statutory Undertaker then that Undertaker must be given not just the powers but the means to fulfil these expectations itself, and not be reduced to begging or hoping that a Local Authority will take over and assist.

Similarly, perhaps, Conservation Areas have been used by Local Government to ensure that attractive canalside environments are protected in some way, but this has often meant a distortion of the planning legislation to fit a linear canal situation. It is worrying for a complete canal through a District to be designated as a long linear Conservation Area (particularly when only BWB property is so defined), and anyway the Board's engineers, whose job it is to operate and maintain the waterways, are often extremely reluctant to agree to Conservation Area Designation because of an implicit fear that their job is going to be made more costly and difficult as a result. Conceived out of the Civic Amenities Act, the Conservation Areas legislation is much more appropriately applied to distinct parts of our historic cities and towns which possess a spatial entity. The canal scene is not quite like this.

In order to solve this problem and make what at present are called Conservation Areas something more than just names on a map, and to give some actual meaning and confidence to legislation, I take the view that a completely new 'animal' needs to be introduced, and I have suggested that it be called an *Area of Historic Landscape Value* for which acceptable safeguards could be agreed.

This would be far better than distorting the existing legislation to include negotiated *exceptions* to various important legal responsibilities. I think it would solve, too, the concern I have for the often expressed view that it is only right to conserve the *best* examples of structures and buildings. In an historic landscape, it is often the quite simple recurring ordinary elements that create the overall value and appeal.

Grants

In terms of grants, the British Waterways Board is a 'grant in aid' authority and, as such, several legal interpretations have made the situation extremely confused, as to whether the Board is in turn *precluded* from gaining access to *additional* grant support from several other Agencies, such as the Historic Buildings Council, the Countryside Commission, etc. Marcus Fox M.P. in 1980 made an interesting decision about the Board's eligibility for grant funding, in that he took the view that nothing prevented the Board from being eligible for grant aid directly so as to help look after its old buildings and improve its landscape assets. However, being eligible for grant aid means the owner needs to be able to contribute at least, say, 50 per cent. of the overall cost of the project. Clearly the Board at present is often quite unable to meet even these costs, and in any event, the other thing about eligibility is that the Board does *not* have the staff resources and professional means to make and co-ordinate, and supervise the grant applications, nor the necessary funds to undertake the work! Therefore both of these problems have inhibited the British Waterways Board from taking advantage, as it otherwise might, of grants from outside sources. As a result, through neglect or impoverishment, our waterways heritage is still being constantly devalued by demolition and indifferent attitudes which could have been avoided perhaps with additional grant support, and a more suitable legal framework within which to operate.

Development Control

However well and carefully the Board manages its canals, towing paths and buildings, the quality of the overall waterway environment is still almost wholly dependent on what happens on land *adjacent* to the canal. Often new canalside development fails completely to acknowledge the canals as an amenity. All too often, fears for safety and security or a lack of awareness will result in ugly canal frontages that seriously neglect the opportunity. It is essential, therefore, when waterside Planning Applications are made, that not only should planting and suitable materials be conditionally introduced, but that opportunities should also be taken to examine some of the basic attitudes affecting the design and layout of the proposal. Such a programme can then avoid serious loss of amenity to the waterway corridor; unless we can influence others to look on canals in a positive way, our waterways will be really adversely affected.

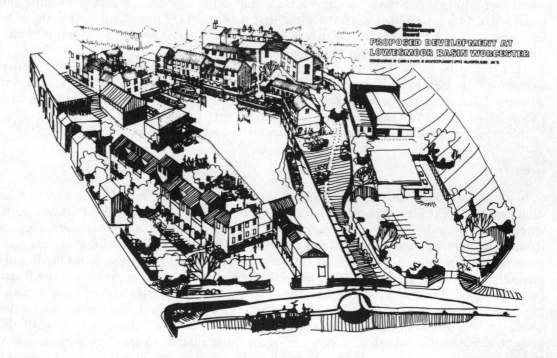

Figure 5.4.   To provide a co-ordinated proposal, somebody must obtain outline planning permission for major development, creating a framework within which the various interested parties can work. This avoids the extremes: dull uniformity, or an uncomfortable rash of ill-matched piecemeal developments.

Even though things like development control and normal maintenance have probably the widest impact on the waterway corridor, the scale and drama of exciting major waterside development schemes is perhaps where most public interest lies.

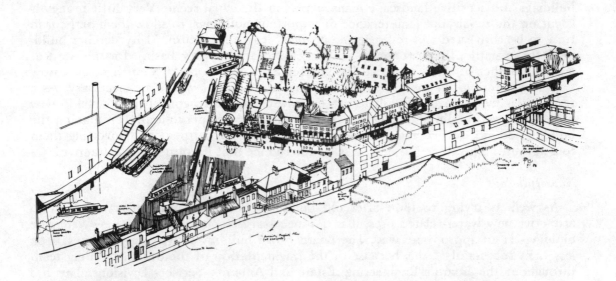

Figure 5.5.   Proposals for Gas Street Basin, designed to co-ordinate individual developments to recreate the character, interest and bustle of the commercial days.

The ownership of non-operational property in urban areas presents to the Board real and tangible opportunities to accrue substantial income by attracting investment and development ideas that can take advantage of the planning gains that water can provide. The focus and sparkle of the canal can be the catalyst which stimulates schemes that include public houses, restaurants, waterside houses, hotels, and shops, as well as canal-based leisure facilities, so revitalising otherwise derelict urban areas. Rather than introduce inappropriate schemes that seek to exploit the canal (so as to gain acceptance that would otherwise be withheld) it is vital to design and plan schemes that can earn money to offset operational deficits, and also schemes that meet the Board's other important duties and obligations, otherwise our waterways' more sensitive areas will be ruined environmentally.

The future must surely lie in marrying old historic buildings and other remnants of our industrial past with well-designed new developments that enhance the historical continuity. For example, at places like Gloucester Docks and Limehouse Basin at one end of the scale, and at equally sensitive sites like Gilwern Wharf on the Monmouth & Brecon Canal at the other, the same attitudes can be demonstrated to produce exciting returns. The recently re-discovered interest in the docklands of London and Liverpool is evidence of the type of scheme that the Board has been trying to progress for some time.

Arrears of Maintenance

Possibly the Board have been too preoccupied, and to some extent understandably so, with attention to the backlog of engineering maintenance as it affects various vital structures of their system and the channel itself. It would certainly seem that dredging and piling works take clear priority over the conservation and restoration of old buildings and sensitive landscape management in the canal scene. Very little money is spent on the repair and maintenance of important buildings, to allow them perhaps in turn to be converted and re-let at an enhanced leasehold return. Many derelict buildings (for example, the warehouses at Chester and at Coventry Basin), for the want of a small number of slates and lead flashing repairs some 15 or 20 years ago, have now seriously deteriorated to the extent that major structural works are necessary. As a result, it is even more difficult to attract private capital to convert and develop new uses for these buildings as substantial sums have to be spent simply on making the buildings sound, secure and weather-tight *before* money is spent on converting them to a new use! This legacy of neglect is an inheritance that is not quite so attractive!

New Buildings

As well as trying to look after old buildings, restoring and converting them to attractive new water-related uses, it is also necessary to plan and design new waterside buildings in an appropriate way. The Board's own building activity has, at times, been less than successful at this because of the fragmentation of the overall building team throughout the Board's Engineering, Estate and Amenity Services Divisions. This has inevitably led to confusion and difficulty, and often a quite disappointing standard of design that hardly demonstrates anything like the developed commitment to design awareness illustrated in the Board's own *Waterway Environment Handbook*! The Board's new control cabins on the Sheffield & South Yorkshire Navigation improvement scheme demonstrate that canal architectural design *can* enhance the personality of a waterway.

But it is *equally* important to look at the 'nuts and bolts' of what is going on *all the time* in the waterway scene so as to prevent unnecessarily crude answers to quite simple and straightforward environmental problems. Success or otherwise may be about the visual impact of galvanised trench sheeting and how it perhaps can be made more visually acceptable by *timber* walings; it might be about whether there is, in fact, any sensible *need* for hydraulic lock gearing; or whether new bridges can be designed in a way that would be more appropriate to the personality of an individual canal; or whether the sensitive historical fabric and the personality of our waterways can in some way be enhanced rather than damaged.

I am sure that it is at this level where the battle for public confidence will be won or lost. I think too that rather than being merely an engineering or political issue it might be about whether or not we can actually manage and conserve an historic landscape. What is the point of dredging canals, introducing cost-effective galvanised bank protection, lopping all the trees back, introducing cheap and sometimes ugly bridges, and making the locks easier to work with hydraulics, if, at the same time as making them A1 in operational terms, our canals have become tedious, boring, sterile places that are seriously and visually depressing? Who is going to enjoy them?

Yes! Of course they have got to be safe and usable — but it must be possible to make them appealing and attractive too. Money isn't the answer in itself — it's what we *do* with the money and whence it can be obtained, which are the real issues.

Conclusion

If I have been able to introduce the view of an architect, so as to influence, stimulate, encourage, persuade and even suggest some answers, then my time with the Board will not have been badly spent. I began by appreciating Hadfield's histories, Rolt's narrative, and Eric de Maré's *The Canals of England*, and towards the end of the latter, as I have pointed out earlier, Gordon Cullen contributed a sketch that made me wonder about the future. It is because they possess such a great past and a challenging present, that our inland waterways deserve a decent future, and this act of our play is only perhaps just beginning ... Let's make sure that there is something *left* worth caring about — for us to pass on to the next generation. Look around you!

Note: The author is Architect/Planner to the British Waterways Board, and the views expressed in this chapter are his own personal views, and not necessarily those of the Board.

CHAPTER 6

WATERWAY RESTORATION: PUBLIC MONEY, PRIVATE MUSCLE

DURING the past two decades public attitudes towards the relics of the industrial past have changed dramatically. The shorter working week and sometimes enforced unemployment have led to a greater interest in the use of leisure time and more particularly of leisure space. One way in which these changes can be identified is through the ever-growing number of inland waterways which are currently being, or have already been, restored. The history of the waterway restoration movement has been recorded by the author in *Canals Revived* (Moonraker Press, 1979) and *The New Navvies* (Phillimore, 1983) but these tell only one part of the complex story. There are in fact three major elements in the revival process and each must play its part for the new look to be achieved. First and foremost is the political campaign. The late Robert Aickman, the founder of the modern revival campaign, always used to say that it took 20 years to change public opinion. Once you had changed public opinion — 'got the politics right' — the first hurdle was overcome. Thereafter money and labour were the other factors which had to be co-ordinated for the revival plans to achieve complete success.[1]

In this day and age, finance looms large on all horizons. It equally tends to dominate the waterway revival scene. Little can be achieved without the appropriate financial backing. This chapter therefore examines the way in which the financial key has to be found to unlock various closed canals. It also offers an opportunity to examine the various ways in which the variety of different problems which had beset a network of uncoordinated and derelict inland waterways have now been overcome.

There are perhaps five main components to the financial equation, each of which can play a role, to a varying extent, depending on the original legal base on which the waterway was defined. Often the combine to achieve the desired aim. These elements are:

1. Charity
2. Government aid
3. County Council support
4. Industrial support
5. Individual donations.

Each offers a different route of access to funds, but all ultimately result in the achievement of a single aim — the revived waterway.

Perhaps the most fundamental advance in relation to finance was the recognition of the possibility of designating as registered charities groups working towards the revival of waterways. This not only gave them an air of respectability, but offered considerable tax advantages as well. The first breakthrough came in 1949 when the solicitors for the embryo Lower Avon Navigation Trust devised a scheme whereby the navigation could be restored by a limited liability, non-profitmaking, charitable undertaking which not only could secure the best political terms, but equally could gain substantial tax relief on all its funds.[2] Such a facility provided both the means to stimulate public confidence and a way to maximise the value gained from hard-won cash. The success of the Lower Avon revival is well known. It took the Trust some 14 years to complete its goal. During this time it collected and spent £81,444 on the refurbishment of the navigation, which would otherwise have fallen into greater decay.[3]

The concept of waterways as charities is not new. In fact, the Driffield Navigation was originally constructed as a charitable concern.[4] However, the prime aim of the majority of the original waterway promoters was profit. Waterway schemes were devised to provide a commercial service for their locality, and at the same time offer a reasonable return for their shareholders. Once the original purpose had been fulfilled, the normal economic forces should have ensured that their demise was as swift as their growth. However, waterways were different insofar as the majority became, under statute, 'water highways' which could not, normally, be closed unless it could be proven that they had not been used by the public for at least three years. It was in fact the proof of use, however occasional, which enabled many canals to stay alive, even though their original commercial role had long since passed. Many canal Acts also provided the right of pleasure use, often without charge, so long as locks were not used. This gave an early rise to leisure traffic and has since provided the means for many of them to remain alive.

The fact that canals have now become living relics of the industrial past has meant that most of the inland navigation network is rightly deemed part of our national heritage. It was through this heritage link that the National Trust, in 1960, was persuaded to take over part of the Stratford-on-Avon Canal, one of the most picturesque rural canals, which then lay derelict, and to use its resources, as one of the foremost national charitable organisations involved in the protection of the nation's heritage, to promote the rejuvenation of the canal. Although the official estimate for re-opening the 13-mile waterway was over £135,000, the National Trust set itself the task of reviving the waterway for £42,000, some £22,000 of which it had to raise from public donations, with the balance coming from a contribution from the Ministry of Transport. The actual cost of the restoration, when the canal was re-opened in 1964, was £53,000 – of which £10,000 had been donated by the Pilgrim Trust and some £23,000 by the public. The proven status of the National Trust offered the means to get much of the physical work undertaken by volunteers, although prisoners from Winson Green Jail, together with servicemen from local Royal Engineers and Royal Air Force units, all played their part.[5] The restored canal was formally taken over by the National Trust in 1965 and its future then seemed assured. However, in 1980 the Trust indicated that it could no longer bear the burden of maintaining the canal, for which the income £27,000 (1979) was far exceeded by the expenditure of £52,000. The National Trust wanted to hand over the canal to 'a suitable body' by the end of 1981. Local enthusiasts, however, had different ideas and suggested that the way out of the crisis would be for them to form a new charitable Stratford Canal Trust, which could take over the canal and re-restore it, by using volunteer help, for even greater leisure use.[6] The cycle thus repeated itself.

The original revival of the southern section of the Stratford Canal brought to light one of the first methods by which Government money could be used to stimulate the revival of a canal. In February 1959, the Government published a White Paper setting out their proposals for the future of inland waterways. The White Paper indicated 'the Government would be prepared, in principle, to bridge a small gap by a special *ad hoc* grant towards the capital cost of redevelopment'.[7] Before this gesture, Government financial support for waterway revival projects had been unknown. It did, however, set a precedent, which has been continually expanded ever since.

Perhaps one of the earliest and most constructive developments came through the work of the Government-sponsored Inland Waterways Redevelopment Advisory Committee. This suggested, in respect of the Kennet & Avon Canal, that basic funds allocated annually for maintenance by the British Transport Commission (BTC) might be augmented by fund-raising and voluntary efforts by the Kennet & Avon Canal Trust Limited, which was incorporated in 1962 and registered as a charity. This principle of the use of voluntary help, complemented by charitable funds, was accepted by the BTC's successors, the British Waterways Board (BWB), and from 1964 the BWB and the Kennet & Avon Canal Trust Limited became active partners in the revival of that waterway. Between 1965 and 1981 the Trust raised nearly three-quarters of a million pounds for restoration work.[8] With this, it has procured the restoration of some 43 locks, the re-lining of the canal between Limpley Stoke and Avoncliff, and the restoration of two old water pumps at Claverton and Crofton. The aim of the Trust is now firmly set on the complete re-opening of the whole 86-mile waterway by 1990.

It is perhaps worth considering how rapid the momentum of the Kennet & Avon Canal Trust's progress has now become, mainly due to its proven ability to raise funds. In the 13 years from 1962 to 1975 the Trust raised £238,149 from grants, membership fees and donations, and by 30 November 1975 it had spent £214,362 on the restoration of the canal.[9] In the 20 years however, from 1962 to 1982, its total income had risen to £834,648 and of this it had been able to spend £672,309 on restoration and had made commitments with BWB to procure further lock restoration at Padworth and Aldermaston in 1982 and 1983. A detailed study of the Trust's accounts also shows that a variety of charitable trusts, including the W. A. Cadbury Trust, Robinson Charitable Trust, Boots Charitable Trust, Manfield Trust, Englefield Trust, Frognall Trust, Pilgrim Trust, Harris Trust, Robin Hill Trust and the Paling Trust, had all made donations towards its work.

The restoration work on the Kennet & Avon is by far the most significant example of the way in which charitable donations are helping to revive a nationalised waterway. In this instance the Trust are able to negotiate with BWB about the way the restoration should evolve, and then pay for it to be undertaken by the Board's own staff. In other instances of canal restoration, a more direct involvement in the work has developed. This will be considered later in the context of the revival of the Welsh Frankton Locks on the Montgomery arm of the Shropshire Union Canal. However, before we leave the question of partial Government finance for restoration works, there are other aspects which need to be considered.

Some of these derive from the 1968 Transport Act, which was the result of a Government review of the future role of canals, the second within a decade. The thinking behind the 1968 Act was outlined in a White Paper (Cmnd 3401) *British Waterways: recreation and amenity*. Amongst other things it suggested that partial finance for restoration schemes might come from sources other than the waterways budget, especially where local amenity was concerned. The resulting 1968 Transport Act split the nationalised inland waterways into three categories: (1) commercial, (2) cruiseways, which were given some security as recreational waterways, and (3) 'the remainder'. Section 107 of the Act laid down a duty on the Board that each waterway in this last category should be dealt with in the most economical manner possible (consistent, in the case of a waterway which is retained, with the requirements of public health and the preservation of amenity and safety). The Act also enabled BWB to seek out and negotiate financial agreements with local authorities and other similar bodies, for the revival of their local canals.

A good example of the effect of this ability to 'tap' additional funds is provided by the Caldon Canal. In this instance, the Staffordshire County Council and Stoke-on-Trent City Council were interested in seeing the canal restored, both as a recreational amenity and perhaps, in the longer term, as a tourist attraction. After detailed negotiations with the Board's staff, the two Councils agreed in November 1971, to contribute to the restoration of the main line of the branch canal at an estimated cost of £85,000. Of this, £50,000 was to be provided by the two councils, partly assisted by a Department of the Environment Derelict Land Restoration Grant, whilst BWB agreed to provide the balance of £35,000.[10]

Many other canals have since been restored by BWB in much the same way. It is however worthwhile mentioning another notable example in the Titford Branch of the Birmingham Canal Navigations. In this case the imminence of local government reorganisation provided much of the driving force to allow local finance to be used to revive the waterway that leads to the highest level of the BCN system. In this instance the Borough of Warley, which was due to be abolished under local government reorganisation, took the initiative, and proposed a scheme which would enable it to integrate the then derelict Titford Pools area into a new park, which they planned to create from adjacent derelict land already owned by the Council. For this scheme, the Council worked with BWB to complete the £20,000 revival plan, and also contributed a further £3,000 towards the restoration of the Oldbury locks.[11]

Derelict Land Grants were not the only Government funds available to assist waterway revival. In fact, a more fruitful source for one scheme was the money available under the various Land Drainage Grants and, subsequently, the Amenity Budget of the local Water Authority. It is certainly the case that the impressive rebuilding of the Great Ouse Navigation between Roxton and Bedford could never have been achieved so magnificently in any other way. (The story of this revival is told in *Waterways World* for May 1978.) Here, the Great Ouse Restoration Society only needed to provide the extra finance necessary to bridge the gap between the cost of essential drainage works and the provision of additional equipment needed for navigational use. For the most part, this meant the provision of an additional set of lock gates.

In 1961 the cost of restoring Cardington Lock was £15,000 and the Society agreed to provide £1,000 to meet the cost of the extra gates. By 1968, when plans were made to rebuild Roxton Weir as a water-measuring station, the Society offered to contribute £5,000 towards the cost of a new lock in the scheme. Society plans for future developments were thrown awry when it was realised that there was likely to be minimal land drainage benefit from a new lock and weir, needed for navigation purposes at Barford, which could cost £140,000 − and as such there could be little help from Drainage Grants. However, in October 1972, Bedford Town Council agreed to contribute £100,000 towards the cost of re-opening the river to Bedford, and this provided the 'key' to this particular crisis. Again, the Society offered to raise and contribute £5,000 towards the cost of the new lock and, to bridge the remaining gap, offered a supply of volunteer labour which could be used to reduce overall costs. Subsequently the Society committed itself to raise £5,000 towards each of the remaining two new locks at Willington and Castle Mills, which were needed to re-open the navigation to Bedford. At Castle Mills, the total cost of the new lock and dam was some £370,000, because of the need for automatic sluices and the demolition of the old stanch below the lock. The final re-opening of the navigation to Bedford was achieved in May 1978. By this time the Great Ouse Restoration Society had raised £22,000 to finance the navigational works out of a total investment of some £750,000 for the complete revival scheme, the balance being funded by the Drainage Authority, assisted by Government Grants and the magnificent gesture by the Bedford Town Council.[12]

Various other central Government funds also have, indirectly, contributed towards restoration works. The Countryside Commission and the National Parks Authority both have the facility to assist restoration. Perhaps the most beautiful of these has been the re-opening and subsequent re-reopening of the Brecon & Abergavenny Canal. Much of the length of this delightful waterway is within the boundary of the Brecon Beacons National Park. As early as 1961 it was estimated that it could cost up to £25,000 to put the canal in order and that thereafter there would be an annual maintenance bill of £10,000. It was not until 1968 that an agreement was finally reached between the Monmouthshire and Brecon County Councils that the canal should be restored at a cost of £28,000 and that subsequent maintenance costs should be shared equally between BWB and the County Councils.[13]

In 1975 there was a serious breach on the canal at Llanfoist and it was estimated that it would cost of £50,000 to stabilise the bank. This finance was provided by the National Parks Authority.[14] However, the estimate of £483,000 to re-open the canal completely was a different problem.[15] Fortunately, by 1977, other Government funds were available for relieving unemployment and these were quickly used to reconstruct some 3½ miles of the canal. Under the terms of the agreement for this work, the BWB provided the plant, supervision and expertise, Powys County Council sponsored the labour from Job Creation grants, whilst the Welsh Development Agency provided the materials needed.[16] However, due to other necessary works, it was not until 1980 that the canal was finally re-opened. By then National Parks Authority and Welsh Tourist Board grants had also been used to finance the balance of the scheme.[17]

Perhaps one of the greatest stimuli to waterways revival has been the facility to inject funds, and more particularly labour, through the various Manpower Services Commission Job Creation and subsequent Youth Employment and Work Opportunity schemes. The Job Creation Programme was introduced in October 1975 and had an initial budget of £70 million. By 1977 over 450 people were being funded under it, at a cost of over £600,000, on projects associated with canals. It is very difficult to get a later overview of the variety of schemes since no central register is kept, but an article in *Waterways World* for March 1977 listed some 20 projects where full-time employment on restoration was being achieved at Government expense. Perhaps the most significant of the schemes in that list was a £270,000 project being co-ordinated by the Rochdale Canal Society, employing some 184 local people.

The Rochdale Canal Society, like many other groups, managed to maintain the impetus behind this type of project, and by 1982 was able to tell its members that a further two-year scheme to employ a total of about 140 people at a cost of £120,000 — shared equally between the County Council, Calderdale Council, and the Manpower Services Commission — had been agreed. This latest scheme included the restoration of a further five miles of canal and the renovation of ten locks between Todmorden lock and Hebble End at Hebden Bridge. Similar schemes were also under way in central Manchester and within the Rochdale Borough, where further restoration was being attained on the same canal.[18]

Such Job Creation schemes have also provided the means to overcome what had previously seemed insurmountable problems. Perhaps the best examples of these are to be found on the Kennet & Avon Canal. In 1978 work was able to progress on two of the remaining major barriers to the re-opening of the canal, the dry section at Limpley Stoke, and the flight of 29 locks at Devizes. Under the former, a 1¾-mile section of the canal, which for many years had had a history of leakage, was being completely rebuilt at a cost of £110,000 for materials alone, while the Manpower Services Commission offered labour to the value of £138,000 and BWB contributed plant and fuel valued at £25,000. At Devizes, under a separate Job Creation Scheme carried out by the Kennet District Council and the BWB, nearly all of the restoration work on the lock flight was similarly completed, including the dredging of the side ponds to their original depth and the repair of masonry and brickwork.[19] Tasks of such magnitude would clearly have taken much longer if it had not been for the availability of the Manpower Services Commission finance.

It would be wrong to give the impression that no major restoration hurdle could not be overcome without central Government aid. In many instances County Councils have been in the lead, using ratepayers' money to provide the full range of amenities that a restored canal can offer. Two very different examples of development will demonstrate the range of options that have been seized by various County Councils.

In Devon, the County Council acquired the Grand Western Canal from the BWB in summer 1971. Since that time it has set out to convert it into a linear Country Park. Although the canal has no locks, it did have a dry section and an aqueduct which both needed extensive repair, and the whole 13½-mile line needed substantial clearance and remedial work to make it navigable. The restoration of the aqueduct alone cost over £50,000 as it had to be relined with a waterproof membrane. Similar treatment was given to the dry section at a cost of a further £19,000. Once the canal was re-watered, a horse-drawn trip boat was introduced to ply along its line in 1974. In April 1976, when the Taunton to Exeter motorway was opened, the full value of the Grand Western Canal Country Park, achieved through Devon County Council investment, was available for all to enjoy.[20]

Of a more extensive nature are the roles of the Surrey and Hampshire County Councils in the revival of the Basingstoke Canal. In 1973 these two Councils sought

Compulsory Purchase Orders to acquire the canal. These were confirmed in 1975, but in the event the Hampshire County Council took possession of the western end of the canal in 1973 and Surrey acquired the eastern end by negotiation in March 1976, both Councils using their Amenity Development funds. Since that time the Councils have worked in concert with the Surrey & Hampshire Canal Society, to help it revive the waterway. In 1981 this amounted to a £114,900 budget contribution from Hampshire County Council which was only able to recoup £7,900 in fees for the use of the canal, and a £91,200 contribution from Surrey, of which the County found £40,600 and the riparian District Councils found the balance, which was offset by £10,000 from fees.

At the outset, the Canal Society undertook to provide a constant supply of volunteer labour to assist with the works; this was valued at £50,000 for 1981. Local fund-raising events have since helped the Society to purchase and renovate an old steam dredger, whilst a £1,000 donation from Johnson Wax of Frimley purchased two Bantam tugs and some old gravel barges to speed up the dredging works. In 1981 the Society raised a further £6,072 in donations, and £4,310 from a sponsored walk, and membership fees brought in a further £8,566, the greater part of which was immediately spent on the canal. Both County Councils are employing wardens to develop the canal and have used their influence to facilitate the development of a variety of Job Creation Schemes, especially by guaranteeing adequate supplies of materials and professional expertise.

In 1977 the restoration works received a £52,000 grant from the Manpower Services Commission to employ 26 full-time staff on lock restoration on the Deepcut Flight. In 1978 this was followed up with a further £130,000 grant to employ some 45 full-time staff to enable the restoration to go on apace.[21] This in turn was followed by additional grants totalling over £300,000 to enable the whole lock flight to be restored.[22] In 1981 alone, a Work Experience Scheme was employing six staff and 12 young persons per week, with a total wages bill of £68,272 of which the Manpower Services Commission refunded £51,906.

Local firms were also encouraged to make donations towards the works, and this has proved a valuable source of funds, just as the trip boat, *John Pinkerton*, run by the Canal Society, has also proved a money-raising success. In 1981 the trip boat produced some £11,633 profit towards the restoration works and, with the extension of its cruising routes, similar sums are likely to be achieved in subsequent years.[23] Much of the money raised has gone toward re-gating the locks of the Deepcut Flight, whilst a £5,000 prize won in a Spar groceries competition enabled Society volunteers to start work on restoring the top lock on the St John's flight.[24] The momentum of the work is such that a target date for full re-opening is now set for 1985/6. This is certainly a viable proposition in the light of County Council support, for it is their financial reassurance which has made the whole project take on its air of success.

This story of financial reassurance by Councils is mirrored throughout the restoration scene. Clearly it is impossible to list every scheme, but one perhaps stands out because it concerns a tunnel going under a Borough, rather than a simple canal running through it. The tunnel is of course the Dudley Tunnel, and the Council the Dudley Borough Council, which was also active in promoting the idea of the development of a Black Country Museum, at much the same time as the prospect of the tunnel revival came about. After extensive negotiation between the BWB and Dudley Borough

Council during 1969 and 1970, an agreement was concluded whereby the Council would financially support the restoration of the tunnel and the re-opening of the derelict Lord Ward's Canal through the work of two charitable Trusts. The latter scheme provided the central focus for the designated Black Country Museum site. The tunnel restoration began in earnest in January 1972 when the Dudley Borough Council made an initial donation of £4,075 to the Dudley Canal Trust's Restoration Appeal fund.[25] Ultimately, the Council was to provide over £7,000 towards the overall scheme, and local firms, charitable trusts and private individuals donated over £5,000. The Council's money alone was insufficient to get the project under way, but it did provide the backing to persuade major organisations, like the British Steel Corporation, to donate the free use of equipment to speed up the restoration work. Other organisations offered hire plant at very favourable rates, to enable the overall costs to be contained within the Dudley Canal Trust's budget for the works.[26]

The official re-opening of the Dudley Tunnel took place on 21 April 1973, and from that date the Canal Trust resumed its regular trips through the tunnel so that the public could see for themselves the many points of interest within. Since that time the Trust have introduced electrically-powered boats to cope with the ever-growing volume of visitors who flock to see this unique example of early canal engineering. These boats have, in turn, continued to raise funds for the further restoration of the canal and have assisted the Friends of the Black Country Museum in their task of re-creating those images of the Black Country scene. For this task alone, local industry has contributed more than £600,000 which, together with other grants, has ensured that the museum now preserves facets of Black Country life which would otherwise have been lost.

Councils' commitments to canal restoration have sometimes evolved in a less direct way, but the ultimate effect has been the same. A case in point is the revival and restoration of much of the Neath and Tennant Canals by the local Canals Preservation Society. In this instance, Neath Borough Council were willing to sponsor a Community Canals Officer, who could act as a full-time paid liaison officer for the Society, and who could also organise community-based projects to assist with the re-development of the canal. The Council agreed to meet one quarter of the costs involved, with the balance coming from a five-year Urban Aid Grant funded by the Welsh Development Agency. The grant also covered the basic facilities of transport and administration expenses which enabled the officer to function effectively. Overall, this facility amounted to a total contribution of £12,560 in the 1980–81 financial year, with even more guaranteed in subsequent years to cover inflation.[27] Through the Community Canals Officer the prospects for longer term restoration of the canals were thus able to develop in a far more structured way. This is best shown by the analysis of his duties which include:

1. the establishment of links with agencies involved with special and social priority school pupils, young offenders, and unemployed young people;
2. the liaison for, and the development of, suitable works projects for young people;
3. the development of publicity for the canal project and the necessary materials to educate participants in the historical and natural interest of the canals; and
4. the development of provision for the use of the canals by youth groups for water sports, etc., including the provision of a trip boat for the use of handicapped youngsters.[28]

One should not forget that once there is a Council commitment to the restoration work in this way, many other opportunities for access to public funds also become available. One way in which the Neath & Tennant Canals Preservation Society have tried to develop this to its advantage has been through the preparation of a review of tourist potential — *The Next Five Years*. In this, the Society presented a case for Tourist Board Grants to assist in the revival of the canal by the provision of canal-side walks, restored buildings and the development of a Country Trail. For each of these themes it identified the potential and indicated how this could be achieved.

This same theme is now evolving elsewhere in Wales with similar success. Possibly the best example is on the Montgomery Canal, where an Interpretative Officer has been sponsored jointly by the Prince of Wales' Committee and the Powys County Council, with the support of the Manpower Services Commission and the Carnegie U.K. Trust/ Countryside Commission Interpretative Projects. Here the aim of the initial project was to investigate ways of bringing the whole canal back into the life of the area. This has largely been achieved in three ways. Firstly, there was the need to analyse the potential of the canal and to develop ways to interest and involve many people in it. This was brought about by the organisation of an annual 'Diary of Events', which included local history tours, guided walks, and work on the canal, together with canal rallies, fishing competitions and various 'folk' activities connected with the canal. Secondly, a group of three previously unemployed young people were sponsored by Powys County Council to produce some publicity material to explain the facilities on the canal, and to encourage its use. The result was a series of booklets and leaflets under the theme *About the Montgomery Canal* which detail boating and fishing facilities, as well as offering a full analysis of what is to be seen along the whole canal route. Thirdly, the Interpretative Officer planned to develop a 'centre' for the canal in an empty warehouse in the Canal Yard at Welshpool, where a permanent display and information centre could be created to attract even more interest in the renovation and revival of the Montgomery Canal.[29]

The first restored section of the Montgomery Canal has also been the trial area for one development which has generated a further willingness by various official bodies to consider other restoration plans sympathetically. This has been the innovation of waterway cruises for the handicapped. One of the first of these schemes was promoted by the Prince of Wales' Committee and the Variety Club of Great Britain. Their boat *Heulwen-Sunshine* now travels a restored 5½-mile length of canal, where recent work has re-opened four derelict locks, to offer a trip of up to six hours. The boat, constructed by Birkenhead shipyard apprentices, was the first of its kind, being specially designed to facilitate cruising by up to nine handicapped people together with their helpers. Joan Heap, wife of the Inland Waterways Association (IWA) Chairman, organised the appeal which raised the £5,482 necessary to build the boat, with the help of the ladies of the IWA. The success of *Heulwen-Sunshine* has now been followed elsewhere in Britain, and more special boats are being built, through the Rainbow Boats Trust, to meet the ever-growing need.

Despite all these examples, it is not true to say that restoration can only take place if Government or Council financial backing is available. In many instances it has been the sheer hard work of individuals and societies, in their fund-raising efforts, which has got projects under way. Often these have been on the non-nationalised waterways.

On the State-owned waterways, the BWB have usually sought to retain control, but in one case enthusiasts actually managed to gain BWB permission to restore a section of a State-owned canal, which the State could not afford to renovate itself!

The project in question is the IWA section of the Montgomery Canal. Here, the four Welsh Frankton locks are to be restored by the IWA using volunteers, with the Waterway Recovery Group (WRG) leading the assault. The scheme was conceived jointly between the Shropshire Union Canal Society and the IWA as the only means of getting the derelict linking section of the canal restored. The project is now being co-ordinated by the Montgomery Waterway Restoration Trust, which was formed in 1981 to develop the scheme. This Trust has the active support of the Prince of Wales' Committee, which has substantially enhanced its credibility. Perhaps a measure of the success of the scheme can be derived from the mass of people who attended a ceremony on 10 October 1981 to install the top gate in the first completely restored lock. In the two years that the project had then been underway, volunteers had completely rebuilt two locks and the pound between locks 2 and 3 had been fully restored. Top gates for locks 1 and 3 had been purchased and fitted. The BWB had dredged one mile between Rednall and Queens Head, and had also piled the area around Frankton Junction. By that time some £40,000 had been raised by the enthusiasts, and spent on the restoration works, together with many thousands of free man-hours. The money had been derived from various sources. In particular, the Manchester Branch of the IWA agreed to raise the £5,000 towards the cost of the two new lock gates. This they did by reproducing paintings for sale, as well as organising many other fund-raising events. The WRG also played its part in raising funds by the sale of commemorative medals and the offer for sponsorship of cast iron mile posts, amongst various other money-raising schemes.[30]

Other finance came from the IWA's National Waterways Restoration and Development Fund, which by December 1981 had granted £24,500 to help the Montgomery restoration. That self-same IWA Fund, since its inauguration in 1969, had raised £160,000 towards national waterways restoration aims by 1981, and of this £114,000 had been distributed to a variety of schemes, with additional grants towards plant hire costs and survey fees.

Not all projects are as large as that at Welsh Frankton, yet many are equally as important to preserve the future of local waterway networks. In some cases the financial cost may be counted in hundreds rather than thousands of pounds. A good example is to be found at Well Creek, where the Well Creek Trust, formed by local enthusiasts, has relied on regular subscriptions from members, and larger donations from some others, to keep its local waterway open as a link between the Nene and the Great Ouse.

When the group first started its work in 1972, it relied on volunteer labour to clear over 250 tons of rubbish from the Creek. In 1975, however, actual fund-raising and the permission from the Middle Level Commissioners were required to enable a further 600 cubic yards of spoil and rubbish to be removed from the waterway. Since that time local fund-raising has continued, enabling a boat basin to be re-excavated at Outwell, and the bank sheet piled, as well as new landing stages to be built, and a village staithe to be established at Nordelph. It is only through the active support of local enthusiasts in raising the necessary funds, and the work of the Well Creek Trust, that this vital link waterway has been re-opened and upgraded to its present form.[31]

In other instances, groups have had to raise funds to finance their fight to gain permission to re-open their local waterways, even before large-scale physical restoration work could begin. Probably the saddest case is that of the Yorkshire Derwent Trust, which was set the daunting task of raising £20,000 to meet potential legal fees in a battle to prove that there was a right of navigation on the river which it ultimately hopes to restore. Such was its actual support that when it made a public appeal in April 1981, it had already raised £10,000 and was well on the way to meeting its target of raising the balance by 31 August 1981.[32] Such is the strength of spirit which provokes individuals to fight for what they believe is right! It must also be remarked that even before that legal battle, the group had raised, or been promised, some £17,000 to help re-open the navigation. The group had also been instrumental in re-opening Sutton Lock in 1972 at a cost of some £3,800, and had renovated Howsham Lock and rebuilt Howsham Weir. It had also acquired the land on which to build a new lock at Kirkham Abbey, where the old lock had been destroyed by a water control scheme.

In many ways the Yorkshire Derwent Trust is typical of the many societies who are still fighting to re-open their waterways. As its prospectus so aptly states: 'The restoration of the river above Sutton *will* proceed as quickly as funds become available'.

This same fighting spirit has achieved something which in 1964 many thought would be impossible: the raising of some £350,000 to rebuild the long derelict Upper Avon Navigation, between Evesham and Stratford. However, as with all good fairy stories, sometimes there is a remarkable benefactor who wishes to help, just at the time a scheme is planned. The late Robert Aickman subsequently recalled a conversation he had at about the time the Lower Avon and Stratford canals were re-opened in 1964, when a gentleman remarked to him, 'Now we must restore the Upper Avon'. Aickman replied 'Indeed we must, but the estimated cost is £350,000'. The gentleman then asked, 'Would it make any difference if I offered a hundred thousand pounds?'[33] Well, that certainly got the project under way! But, even before the physical work could start a new Navigation Act was required and this cost some £2,000, together with much effort. After that the real process of fund-raising began. Initially, the £100,000 was to be made available on the basis of £1 for every £2 raised elsewhere. This was later amended to £1 for £1, with the ceiling set at £120,000. To match this, the Upper Avon Navigation Trust managed to raise over £190,000 by December 1973; of this amount £25,000 came from the Sports Council, £12,000 from the IWA and £5,000 from the Countryside Commission. Charitable foundations provided further large sums, and one family, the Billingtons, donated £15,000 on their own. Local authorities provided some £2,000. Careful management enabled the re-opening of limited sections of the river once they had been restored, and boat permits to use these waters raised another £1,500 per annum.[34]

Nearly all the work was completed by October 1973, but £25,000 was still required to clear the balance of the loans made to create the works. As a publicity stunt, one small earth barrier was left across the navigation channel and the restoration manager, David Hutchings, said 'If money comes in we shall open – if not, we shall not' as he made his final appeal.[35] That last dramatic appeal achieved its aim, and the 17-mile waterway was re-opened by the Queen Mother on 1 June 1974.

Unfortunately, a restored waterway needs constant funds thereafter to keep it open and, since re-opening, the Upper Avon Navigation Trust has managed to balance its books by careful management. In 1980, however, it looked as though a further lump sum of £41,000 would need to be raised to carry out repairs to Welford Weir. Fortunately, local boat hire companies donated £3,000 for immediate remedial repairs, and the Trust set about raising funds for a more permanent repair, which was eventually achieved at half the estimated cost.[36]

In 1981 it became evident that maintenance problems and navigational difficulties in the area of the Robert Aickman lock, at Harvington Mill on the Avon, could best be overcome if a new lock and channel were built. By coincidence, land became available to permit a better alignment and the Upper Avon Trust set about raising the £150,000 needed for the scheme. Legal constraints made early decisions essential and, as an act of faith, the Trust commenced its task in advance of its appeal. As part of the appeal effort, the Trust was able to sell the derelict Harvington Mill, and use part of the proceeds as a donation to the lock appeal. The IWA also agreed to promote an Appeal in autumn 1981, as a Memorial Fund to its founder, Robert Aickman, who had died earlier in the year. The aim was to raise some of the cash necessary to provide a true working memorial to his achievements in the waterways world. Such was the momentum of Appeal that over £24,000 had been raised from donations by the end of 1982.

In many ways, one man has been synonymous with the dynamic manner in which the Upper Avon was rebuilt. He is David Hutchings, M.B.E. He has, however, had to work with and through the controlling body of the Upper Avon Trust. Much of the credit must go to his ability and skill in grasping opportunities to move forward as and when the need arose.

Another individual, Alan Picken, acted rather differently, as he believed very much in going it alone. In 1978 the 2¾-mile Coombe Hill Canal in Gloucestershire came up for auction, and Alan Picken decided he would buy it and restore it under his own steam. He was lucky in that the other major bidder dropped out of the auction just before his limit of £35,000 was reached, and at the end of the auction, he was the proud owner of the canal. His aims were two-fold: firstly to restore the wharf buildings and basin at the Canal Head, where a canal museum was planned, and secondly to restore a two-mile section of the canal so that a horse-drawn trip boat could operate along its length. He planned to finance all this himself, through his company, Severn & Canal Carrying Co., utilising the help of volunteers and friends to undertake some of the hard labour of restoring the actual canal. When one considers that some 4,000 tons of silt had to be dredged out from the canal basin alone, it is easy to see what a labour of love restoration work can be.[37] It was unfortunate that cash flow problems finally halted Alan Picken's efforts in 1982, but by then it was envisaged that the newly-formed Coombe Hill Canal Trust could take over the scheme and build on Picken's work. This seems likely to offer new life to a waterway which was abandoned as long ago as 1876, but could still prove a tourist attraction in the years to come.

Finance for the restoration of canals has often been raised on the basis of their tourist potential. A prime example is the restoration of a short length of the Coalport Canal, together with the Hay Incline plane, where the major funding came, in 1973, from a Science Museum grant. This grant enabled the Ironbridge Gorge Museum Trust to purchase the rails and sleepers to re-lay on the incline, and a grant from the Department of the Environment provided funds to allow the Museum Trusts to re-excavate the bottom basin and the top transfer basin.[38]

In many ways the Ironbridge Gorge area can be considered as the birthplace of the Industrial Revolution. It now provides a lasting memorial to the industrial past. Another important site is the canal terminus at Bugsworth Basin on the Peak Forest Canal. This is being restored by the Inland Waterways Protection Society, which has set itself the task of raising all the finance needed to complete the works. The task is such that the group has been working since 1968 to achieve its aims. Much of the work has progressed only as the necessary funds have been raised from donations, awards and grants, but success has led to success. The initial breakthrough came in August 1969 when the first hundred yards were fully restored. This gained the society a Countryside Award in 1970 and gave rise to further donations, which enabled it to continue clearing more derelict areas. The second phase of the project was started in 1975, when the society set out to raise £40,000 to complete the clearance of the Lower and Middle Basins of the canal and tramway interchange port. This part of the project had nearly been completed by 1980, and in 1981 the teams of volunteers started to clear the Upper Basin area. The main problem for the organisers has been the lack of regular funds, but this has, to some extent, been made up for by the enthusiasm of the volunteers and the generous help they have received from some local contractors, both in the loan of equipment and the provision of supplies.

The story of lack of funds for restoration schemes is one that can be told over and over again. It is quite remarkable how some projects have gained regular and adequate funding from the start, whilst others, which seem of equal if not greater merit, have had to fight every inch of the way.

A scheme which falls into the latter category must surely be that for the revival of the Droitwich Canals. This scheme had its origins in the early 1970s, when a sub-committee of the Worcester & Birmingham Canal Society became the forerunner of the Droitwich Canals Trust Limited, which was formed in 1973.[39] Since that time the Trust has set about raising funds for its restoration scheme. It has had the active support of both the Droitwich Development Committee and the Wychavon District Council, which has leased the entire canal line to the Trust for 99 years at a peppercorn rent. At the outset the Trust was confronted with the need to purchase a plot of land at Hawford to allow for the possible construction of a new link into the river Severn, if longer term plans to tunnel under a new dual carriageway which blocked the canal line prove impractical. Fortunately, the IWA were able to lend the Trust £7,000 for this purpose and subsequently were able to convert £6,000 of this into a grant. The balance was funded from donations and an inaugural grant of £1,400 from the Worcester & Birmingham Canal Society.[40] However, as one problem was surmounted so another emerged, in the form of opposition from landowners along the line of the canal. This opposition involved long and time-consuming negotiation, and at the same time substantially retarded the progress of the Trust who by 1976 had only raised £20,000 of the £100,000 needed to complete the scheme. This sum excludes the development of a new basin at the centre of Droitwich, which in 1979 was estimated to add a further £320,000 to the total revival costs.[41]

At that stage, the Trust estimated that the locks alone were costing some £6,500 each to restore. In an effort to raise more funds the Trust had the idea of selling 100 tokens offering free usage of the restored canal to the purchasers, for life, at a cost of £50 each token. The price was subsequently raised to £75 each in 1980.[42]

It was not until the middle of 1980 that the finances of the Trust began to look healthier. One reason for this was an announcement by Wychavon District Council that it would be donating £150,000 to be used in the central area of Droitwich for the canal redevelopment scheme. In addition, this part of the project was also eligible for a £50,000 Department of the Environment derelict land reclamation grant.[43] A little later the Canal Trust heard that some of the new lock gates for the canal would be constructed by Royal Engineers at Chepstow under an Army Apprentice Training Scheme.[44] Even so, the Droitwich Canal Trust still has a large amount of fund-raising to undertake before it finally achieves its goal and, sadly, the re-opening date has moved further into the future.

Not all restoration schemes are such large extended affairs, neither do they involve such huge financial inputs. Some are even co-operative ventures, where small groups of local people band together to finance their own joint revival scheme. A good example of this is the restoration of the very short Hazelstrine Arm of the Staffs & Worcs Canal by the Stafford Boat Club. In this instance, the Club members themselves initially raised funds to purchase an old Ruston Bucyrus dragline crane from the local gravel pit and then used it to clear the arm themselves, raising cash as they slowly progressed. The local council supported their efforts by giving them a lease on the adjacent land.

With this scheme, as the early work began to show results, more people became interested and what had originally been a modest plan to provide a few off-channel moorings grew into a 30-ft. wide basin with piled edge, concrete access road and landscaped surrounds. Some members purchased the steel piling for their own moorings, whilst others raised funds in various ways. Members gave loans to the Club, a local lottery provided £2,000, a small Sports Council grant was obtained, and other cash came from profits from the clubhouse bar. Even the clubhouse was an economy effort, in that it was an old Central Electricity Generating Board site hut, bought by the members for £50. From a tottering start, the Club's plans have since gone from strength to strength, and developments now include a dry dock and a slipway, with the longer term prospect of purchasing further land up the arm to extend the Club moorings in an orderly way. The success of this venture has rested on the continuous flow of new members who were not only willing to finance the work, but also physically to dig out the new berths. This is a practical example of a co-operative restoration scheme.[45]

Whilst some restorations are the work of co-operatives, all restorations require the co-operation of the owners of the navigations or of the land through which the route passes. Sometimes the owners of navigations have a change of heart towards restoration attempts once they see what can be achieved, and subsequently even contribute financially to assist the restoration schemes which previously they had not thought worthwhile.

A good example of this is to be found on the Stroudwater Navigation, which is still privately owned. In 1954 the Proprietors obtained an abandonment Act relieving them of the responsibilities for maintaining bridges and for keeping their waterway navigable, but they still continued to run the waterway as a viable business, deriving proceeds from water supply and drainage rights. However, by 1978, due to the efforts of the local canal society, the attitude of the Proprietors towards the potential revival of the navigation had changed and they started to give the volunteer Stroudwater, Thames & Severn Canal Trust every possible encouragement, including financial aid. It is interesting how times change. Even so, the most passionate supporters of this project accept that full revival must be a very long-term aim, as overall lack of finance precludes dynamic moves.

In the 1981 financial year, the total income of the Trust was £6,487 – of which the greater part came from membership subscriptions of £1,500 and fund raising efforts of £3,149. However, when one considers that the Trust was able that year to raise £2,700 from a sponsored walk, and in 1982 £4,000 from a reception in Gloucester for local firms, it is clear that many local people are willing to support the scheme actively.[46] Conversely, in October 1982, the Trust demonstrated its initiative in attracting funds on a national scale by getting Anthony Burton to make a BBC 'Good Cause' appeal on its behalf. This was the first ever such appeal for a canal restoration project, and not only did it raise £7,498, but also generated offers of equipment and applications for membership. Even with such support, the Trust is still obliged to pursue a step-by-step approach.

The problems encountered by the Stroudwater, Thames & Severn Canal Trust are similar to those of the various other revival schemes, where full restoration to through navigation can only be a long-term aim. Such schemes simply do not have the same potential to attract major financial backing as those where boats are likely to be able to ply within a short space of time. This means that, of necessity, their financial targets are set much lower and their work programmes must evolve in a piecemeal way. Projects are undertaken as circumstances allow, and fund-raising is geared to the gradual approach.

This study would not be complete unless some examples of these long-term schemes were included, even though their works to date are usually on a much smaller scale than the majority of schemes for waterways which have been revived.

The River Stour Trust was set up in 1968 to protect and enhance the right of navigation on the Stour. Since its inception, it has seized every opportunity to revive parts of the derelict navigation works. When the chance arose in 1974 to restore Flatford Lock the Trust immediately instituted a Flatford Lock Appeal to raise the £500 required for the work, as well as organising volunteer working parties to complete the restoration scheme. Local organisations and suppliers rallied round and lent equipment or provided the much needed components at advantageous rates.

A second appeal for £2,000 was instituted when Flatford Lock was re-opened in March 1975. This was to finance a much larger subsequent scheme for the restoration of Stratford St Mary Lock.[47] However, this plan had to be abandoned in 1976 when the Anglian Water Authority rejected the Trust's revival proposals. Fortunately the Trust was able to maintain momentum and organised a Job Creation Scheme, which lasted some 11 months in 1977, to restore a River Stour lighter, which it had previously raised.[48] This project cost £9,322 in wages to employ five men and was funded by the Manpower Services Commission. The remaining £1,000 required for materials was found by the Trust with the assistance of a £500 grant from the Maritime Museum.[49]

A second £40,000 Job Creation Scheme was organised in 1979, which employed nine men for a year, to restore the infilled Sudbury Navigation Basin, and to provide a longer term mooring for the restored lighter.[50] This project had a triple advantage for the Trust. Firstly, it provided a new focus for its longer term aims of generating interest in restoring the whole navigation, especially as a warehouse on the basin side was being re-developed as a theatre for the town of Sudbury. Secondly, it gave stimulus to re-activate the remainder of the former basin area as a larger mooring for boats, and thirdly, it prompted the Trust to consider building a new top lock of the navigation, at Great Cornard, to facilitate the development of a trip boat along the upper reaches of the river.[51]

Fortune seemed to smile on the Trust, and in summer 1981 it managed to get the remaining derelict Gasworks Arm of Sudbury Basin fully restored. On this occasion the project, which would have cost the Trust upwards of £15,000 by normal contract methods, was undertaken free of charge as a training exercise by the U.S. Air Force 819th Civil Engineering Squadron. This involved the removal of many tons of silt and rubble to clear the 100 yards of the basin. It also provided the Trust with access to a derelict granary, which it planned to acquire and renovate for use as a Stour Navigation Museum and headquarters.[52] Such was its progress, that by 1982 the Trust felt able to launch an appeal for £50,000 to construct a new lock at Cornard to circumnavigate the fixed concrete weir built in the 1950s. By the close of 1981 well over £1,000 had been donated and this was doubled overnight with the addition of a further £1,000 grant from the Shell Inland Waterways Restoration Awards scheme.[53]

Just as the Stour Trust was able to make headway with its longer term plans by the adoption of the piecemeal approach, so the Wey & Arun Canal Trust was able to make progress in restoring sections of their canal, and at the same time gain public goodwill by the careful selection of the projects it undertook. Fund-raising was achieved by three main schemes: firstly, an annual sponsored walk, through which in 1981 over 500 walkers raised over £6,000; secondly, a recycling campaign, mainly based on waste paper collection, which raised around £1,800 in 1981; and thirdly, through the sale of booklets and other canal wares. These not only brought in profit but gained general public interest in the canal. Bridge rebuilding and replacement has become a speciality of the Trust. Achievements so far have included the refurbishment of Pallingham and Cooks Bridges, together with the restoration of Rowner Lock and Malham Lock, the rebuilding of brick arched bridges at Loves Bridge and Rowner Lock and, most spectacularly, the construction of a new lift bridge at Northlands. This bridge alone cost the Trust £8,000, even though all the construction work was carried out by volunteers, but it showed everyone just what could be achieved.[54]

On a much smaller scale, but of equal importance, are the efforts of the Wilts & Berks Canal Amenity Group. Whilst the Group accepts that the full revival of its canal is a nearly impossible task, it believes that many sections remaining are worthy of preservation and improvement. To this end it uses volunteers to help to ensure canal features are maintained. Because of this strategy its fund-raising is at a much lower level, being mainly geared to raising money for materials and tools for clearance and renovation work. Much of the finance comes from an active sales stand, which is taken to major waterway events, and other sums come from donations and membership fees. One of the largest amounts was an award of £275 from the 1981 Shell Inland Waterways Restoration scheme, to assist with the renovation of the canal and a bridge at Calne, where the parapet of the bridge needed replacing and the whole structure made sound. It formed part of a longer term project to revive a section of the Calne Branch as a local amenity facility.[55]

Apart from the major fund-raising schemes of the IWA and various individual societies, perhaps one of the most useful financial aids to many revival plans has been the donations from the Shell Inland Waterways Restoration Awards. Three schemes have been launched so far, and overall they have injected some £51,000 in cash to assist various projects. The aim of the schemes, which were organised with the assistance of the IWA, was the encouragement of the organisation of projects, rather than the provision of complete finance.

The first round was promoted in 1977 and £6,000 was distributed to various groups. Recipients in 1978 included the Stroudwater, Thames & Severn Canal Trust, who gained £100 towards the restoration of the Sapperton Tunnel portals; the Nottingham Canal Society, who gained £200 towards restoring two bridges; the Erewash Canal Preservation and Development Association, who received £400 towards work on Langley Mill Basin, and, amongst many others, the Torfaen Canal Society, who were given £60 towards dredging a section of the Monmouth Canal.

The second round of Shell Awards was organised in 1978–79 and on this occasion £20,000 was distributed. Awards included £1,500 to assist the Seagull Trust in restoring a boating base on the Union Canal at Port Calder in Edinburgh, and converting it for the use of the disabled; £1,000 to the Droitwich Canals Trust to enable it to rebuild a Brindley circular weir; £600 to the Inland Waterways Protection Society to purchase a mechanical excavator to speed up the Bugsworth Basin Scheme, and £450 to the Driffield Navigation Amenities Association to rebuild a swing bridge which had become fixed across the waterway.

In 1980–81 the Shell Awards topped £25,000. Amongst the grants made in this round were £1,000 to the Sleaford Navigation Society towards new gates for the bottom lock on the Sleaford Navigation; £1,100 to the Huddersfield Canal Society to help restore a lock and inaugurate boat trips along a restored section of the waterway; £1,000 to the Pocklington Canal Amenity Society to help restore a swing bridge, and so provide access to the Melbourne Basin as a new temporary head of navigation; £1,000 to the Forth & Clyde Canal Society, to assist with its plan to inaugurate a regular trip boat service along a section of the canal; and £700 for the Cromford Canal Society to establish a working museum on the terminal wharf of the canal.

In 1981, Shell decided to develop the theme of volunteer self-help still further — especially amongst young people — and undertook an extensive advertising campaign under the umbrella of the Shell Community Affairs projects and its 'Better Britain' Competition. Full-page advertisements were placed in the national press at regular intervals under the headline 'Fancy helping weed a river?' — with the picture of a canal boat moving along a restored stretch of canal. It offered to put would-be volunteers in touch with local organisations who could use their help. Such advertising offered a tremendous boost to the whole restoration movement and was, of course, at no cost to the individual groups themselves. Such aid 'in kind' is as valuable as the cash itself.

In such a review as this, one clearly cannot cover every facet of finance, but it is perhaps useful to remember that finance for restoration schemes has been derived in a variety of ways, not the least of which were the generous donations by both large and small firms, in cash and 'in kind'. Multinational organisations such as Rank Xerox have provided funds for locks; machinery manufacturers like JCB have provided equipment at less than cost. Insurance companies and building societies have sponsored events aimed at the same basic theme of reviving canals.

*   *   *

It is perhaps useful to reflect upon what all this fund-raising effort achieves. In 1980 the Inland Waterways Amenity Advisory Council completed a review of the benefits offered by canals and presented its report, *Inland Waterways: Arteries for Employment and Spending*. It concluded that in 1979 alone people spent £55 million to enjoy 2,000 miles of canal and river systems controlled by BWB, and that this network supported 14,000 jobs throughout the community, particularly in rural areas. On this basis, each single mile of restored waterway can generate £27,500 of additional trade annually (at 1979 prices) and create an additional seven jobs, apart from the substantial amount of recreation and leisure space that is brought back into active use, or the enjoyment that the volunteers can gain through rebuilding their local canal.[56]

Canals are a living part of our heritage, with an important multi-functional role. The whole community, therefore, benefits from the vision, enterprise and energy of the many individuals throughout England, Wales and Scotland who are working — by fund-raising, by direct action, or both — to give their local waterways a new lease of life. These individuals labour for love, with no thought of gain beyond the satisfaction of contributing materially to the well-being of something they hold dear. Let us salute their enterprise, offer thanks for their achievements, and provide them every encouragement in their task of giving canals 'A New Look'.

REFERENCES

1. HANDFORD, M. The Montgomery Canal – a personal view. *Cuttings* (Shropshire Union Canal Soc. magazine). 63. 1981.
2. *Inland Waterways Assn. Bulletin.* 23. December 1949. 7.
3. LOWER AVON NAVN. TRUST. *Gateway to the Avon.* 1975.
4. 7 Geo. III cap. 97 (1767). *See also* SQUIRES, R. *The Driffield Navigation guide.* 1975.
5. RANSOM, P. J. G. *Waterways restored.* 1974. 48.
6. *Waterways World.* January 1982. 30.
7. DEPT. OF TRANSPORT. *Government proposals following the report of the Committee of Inquiry.* Cmnd. 676. February 1959. 6.
8. *The Butty* (Kennet & Avon Canal Trust magazine). December 1981. 5.
9. KENNET & AVON CANAL TRUST. *Accounts to 30 November 1975.*
10. *Caldon News* (Caldon Canal Soc. magazine). 8 December 1971.
11. *Boundary Post* (Birmingham Canal Navn. Soc. magazine). Titford Canal Restoration Rally Commemorative edn. 1974.
12. *Waterways World.* May 1978. 36–40.
13. *IWA Bull.* 86. January 1969. 65.
14. *Waterways News* (British Waterways Board newspaper). 54. January 1976. 6.
15. *Waterways World.* July 1977. 22.
16. *Waterways World.* November 1977. 31.
17. BRITISH WATERWAYS BOARD. *A plan for the Monmouthshire and Brecon Canal.* 1981.
18. *Rochdale News* (Rochdale Canal Soc. magazine). 25. January 1981.
19. Kennet & Avon Canal Trust leaflet. August 1981.
20. *IWA Waterways.* August 1981. 16–17.
21. *Waterways World.* July 1978. 58.
22. *Waterways World.* May 1980. 56.
23. *Waterways World.* May 1980. 57.
24. *Waterways World.* September 1981. 41.
25. *Broadsheet* (Staffs & Worcs Canal Soc. magazine). 167. March 1973. 4.
26. DUDLEY CANAL TRUST. *Dudley Tunnel: TRAD 1973.* (Commemorative booklet). 1973. 36.
27. *Aqueduct* (Neath & Tennant Canal Soc. magazine). 6. Autumn 1980. 15.
28. *Aqueduct.* 5. Winter 1979–80. 13.
29. GUEST, A. Personal communication. 26 January 1982.
30. *IWA Waterways.* 133. December 1981. 13, 17.
31. WELL CREEK TRUST. *A waterway reborn.* 1976.
32. *Waterways World.* April 1981. 36.
33. *IWA Bulletin.* 110. September 1974. 13.
34. RANSOM, P. J. G. *op. cit..* 57.
35. *Waterways World.* December 1973. 38.
36. *Canal and Riverboat.* November 1980. 25.
37. *Waterways World.* May 1980. 29.
38. SMITH, S. B. Personal communication. 5 November 1974.
39. HODGES, H. *A history of the Worcester and Birmingham Canal Society, 1969 to 1978.* 1979. 39.
40. *Waterways World.* August 1973. 36.
41. *Waterways World.* October 1979. 34.
42. *Waterways World.* December 1979. 24.
43. *Waterways World.* February 1980. 24.
44. *Waterways World.* June 1980. 26.
45. *Waterways World.* March 1982. 25.
46. *Trow* (Stroudwater, Thames & Severn Canal Trust magazine). 35. September 1981. 10.
47. *Waterways World.* November 1975. 39.
48. *Waterways World.* June 1976. 23.

49. *Lock Lintel* (River Stour Trust magazine). 59. Summer 1980. 24.
50. *Canal and Riverboat*. November 1980. 5.
51. *Waterways World*. February 1981. 50–53.
52. *Waterways World*. December 1981. 26.
53. *Lock Lintel*. 65. Winter 1981. 6.
54. *Waterways World*. April 1981. 30.
55. *Dragon-fly* (Wilts & Berks Amenity Canal Grp. magazine). 15. November 1981. 4.
56. INLAND WATERWAYS AMENITY ADVISORY COUNCIL. *News Release 80/64*. 6 August 1980.

A BIBLIOGRAPHY OF BRITISH CANALS, 1623-1950

Introduction

ALTHOUGH THE LITERATURE of railways has for some time been the subject of particular study (Ottley's painstaking work[1] being the outstanding example) canals have not been so well served. Some brief bibliographies have appeared[2] but these have not been extensive either in scope or in detail. It is hoped, therefore, that this bibliography will prove useful, not only to those pursuing research on waterway topics, but also to librarians, book collectors and all those interested in waterway books in a personal or professional capacity.

The unattainable yet perpetually tantalising goal of the bibliographer is completeness. It must therefore be freely admitted that, although nothing answering to the prescribed terms of reference has been deliberately omitted, no claim is made for completeness. The search has been extensive but not exhaustive; this is probably particularly true for books on the Thames and the Broads, both of which have prompted a large volume of literature in the past hundred years, and the pamphlets of the eighteenth and nineteenth centuries, which are numerous and widely scattered. (In the case of the anonymous pamphlets listed in the Appendix, completeness has not been attempted. The aim has, instead, been to provide a representative sample of the genre.)

The last twenty or thirty years have seen a sharp increase in interest in waterways, and thus in the numbers of books published in this field. Many are excellent, some trivial or derivative, a few poor, but most are reasonably accessible in, for instance, the specialist collections established under regional library schemes. By contrast, earlier books are rarer and the mere fact of their publication is itself of interest. Rather than waste time (both writer's and readers') describing modern books of little significance, it was deemed appropriate to be selective largely on the basis of date alone. The terminal date chosen is 1950, as the *British National Bibliography* was started in that year. By chance, 1950 can also be described as the threshold of the modern age of canal publishing as four important books appeared in that year: DE MARE *The canals of England*, HADFIELD *British canals: an illustrated history*, O'CONNOR *Canals, barges and people*, and ROLT *The inland waterways of England*. These provide ample excuse for the overlap of twelve months with *BNB*.

The main list therefore includes books and pamphlets written in English relating wholly or in part to waterways within the British Isles, first published in Britain in or before 1950. Various categories of material, otherwise falling within this general description, have been excluded. Amongst these are Acts of Parliament, and Government and Parliamentary Reports, which can be found in the Official Publications Library (previously the State Papers Room) of the British Library, Bloomsbury, or in the House of Lords Record Office. Company circulars, statements and reports are likewise omitted; reference should instead be made to the Public Record Office at Kew which now holds the main collection of the historical records assembled by the British Transport Commission. These had been split between London, Edinburgh and York, but in 1969 the material in Edinburgh was transferred to the Scottish Record Office. At Kew may now be found all the records previously held at BTC's archives at 66 Porchester Road, London, together with such material which had been at York, and was not duplicated in the London collection.

No attempt has been made to include any of the early engineers' printed reports, partly because of the difficulty of tracking them down, and partly because there already exists a comprehensive bibliography of what must be the nation's largest collection, that of the Institution of Civil Engineers.[3]

These excluded categories can be thought of as comprising the official literature of British canals; this bibliography is devoted to the unofficial literature, that which records the interaction between the public and the waterways. In this interaction, fiction plays a part and so is included, as are a few company publications aimed at a wider readership than shareholders, company officials or existing customers.

A bibliographic description of the first edition of each book is given, following the guidelines contained in the relevant British Standard.[4] In addition, some information is given, where available, of the subsequent publication history of the book. Here, especially, there is no systematic way in which completeness can be attempted, let alone achieved. Nevertheless, it is hoped that such notes as have been provided will prove useful as they give some idea of the original success achieved by the book, and also of its subsequent likely availability.

REFERENCES

1. OTTLEY, George. *A bibliography of British railway history* (Allen & Unwin, 1965). My approach to compiling and presenting a waterway bibliography has been assisted by a study of Ottley's *Introduction*.
2. *See*, for example, the bibliographies in successive editions of HADFIELD, Charles *British canals*, and in PAGET-TOMLINSON, Edward *The complete book of canal and river navigations* (Waine Research, 1978). Other bibliographies have been produced by librarians, e.g. LIBRARY ASSOCIATION Public Libraries Group *Reader's Guide No. 7: Inland waterways* (LA, 1975). Numerous inland waterway items are included within the lengthy but unclassified bibliography in JACKMAN, W. T. *The development of transportation in modern England* (Cambridge University Press, 1916). This contains little twentieth-century material.
3. SKEMPTON, A. W. *Early printed reports and maps (1665–1850) in the library of the Institution of Civil Engineers [London].*|(ICE, 1977).
4. BRITISH STANDARDS INSTITUTION. *Recommendations: bibliographical references.* 1976. BS 1629.

Classification

The material included within this bibliography is presented in the following sections:

1. GENERAL, INCLUDING LAW AND HISTORY. Encyclopedias and similar general reference works are excluded. Works with international coverage of waterways, and works on the constituent countries of the British Isles, are included here, but anything only covering a smaller geographical area is listed in section 2. A few books whose general introductory remarks pave the way for specific proposals are included in section 3.

2. HISTORIES AND GENERAL ACCOUNTS OF PARTICULAR AREAS OR WATERWAYS. Works on whole countries are listed in section 1. Engineering or cruising on particular waterways or in particular areas are placed in section 4 or 6 respectively. General works on un-navigable rivers are excluded. Works on navigable rivers which contain little information on navigation are only included if they describe little-known rivers, or are early works which may contain useful illustrations of locks, barges, etc. A number of books on the Thames, containing nothing of value on inland navigation, have been omitted.

3. PLANS AND PROPOSALS. Excludes engineering reports *per se*, and engineers' comments on engineering reports, although the line between these and the published 'Letters' and 'Observations' of engineers is finely drawn. The aim has been to include only those items which were addressed directly to the public.

4. WATERWAY ENGINEERING; ENGINEERING BIOGRAPHY. Omits works whose sole waterway content relates to the measuring of earthworks.

5. BOATS AND BOATMEN; BOATBUILDING; BOAT PROPULSION AND HAULAGE.

6. CRUISING; CRUISING GUIDES; PACKET BOAT TRIPS. Omits books on sailing technique, building canoes etc., and sailing charts and directions.

7. FICTION. Subdivided into three classes: adult fiction; literature stimulated by the work of George Smith; other children's fiction.

APPENDIX. A selection of anonymous tracts and pamphlets, published before 1840. A good list of such material has been provided by Jackman (*see* Ref. 2 above). Later anonymous works are included in the main body of the bibliography.

Bibliographic information

Information for each entry is given in this form:

Author's or authors' names as shown on the title page, or as signatories to the preface. Extra detail from other sources is shown in square brackets. Names are followed by *dates*, where known. Anonymous and pseudonymous publications are listed under the first word of the title, discounting 'A' and 'The'. If the author is known, a cross-reference is also given, except in the Appendix, which consists exclusively of anonymous and pseudonymous material.

Title of work as shown on the title page. If the title on the cover is significantly different, this is noted.

Place of publication: publisher (or printer), date. Square brackets enclose information derived from elsewhere.

Pagination and illustrations. Square brackets are used where the originals are un-numbered. All illustrations, however reproduced, are described as 'plates' if the pages on which they are printed are not numbered in the main sequence. Information on illustrations given *after* semi-colon means that these have been included within the collation given before the semi-colon (except map endpapers). Blank pages are included only if they occur within a group of leaves which is itself being enumerated. Although publishers' catalogues may vary between otherwise identical copies of books, no attempt has been made here to establish priority of issue.

Brief details of later editions, if any. Usually the latest located is noted. Foreign editions have been omitted. Greater detail is provided for the more important works.

Explanatory comments, where necessary.

Locations and sources

A location or source, but generally only one, is given for each entry. The list below explains the abbreviations used for locations and sources, and is given in order of priority, i.e. a source other than the British Library is usually only cited if no copy has been located in any other source higher up the list. Information from the sources asterisked in the list has not been checked by inspection of the actual copy.

  B   British Library
  S   Science Museum Library, London
  T   Chartered Institute of Transport, London
  C   Institution of Civil Engineers, London
  L   Leicester University
  G   Goldsmiths' Library, University of London
  P   Private collection
*N   National Union Catalogue (U.S.A.)
*(B) In British Library catalogue, but copy known to be missing or destroyed.

To save time, many items were inspected at locations other than the British Library, whose catalogue was then checked to discover whether the items were listed. If they were listed, then the entry in this bibliography will read 'B' even if the British Library copy was not itself actually seen. As the British Library catalogue still lists many items which have in fact been mislaid or destroyed, there will doubtless be a few entries in this bibliography which read 'B' but which should in reality read '(B)'.

During the course of the research, some references were discovered to works, copies of which have subsequently proved impossible to locate. If, in addition, no reliable catalogue entry exists, then such works have been omitted. Doubtless many of the references were wrongly cited initially, or referred to periodical articles or parts of longer works. However, amongst the few dozen rejected for lack of evidence, a few may eventually prove to have been wrongly omitted; their loss is regretted, but is essential to preserve the reliability of this bibliography.

Abbreviations

ad./s.	advertisement/s	facs.	facsimile
attrib.	attributed	fldg.	folding
b.	born	frontis.	frontispiece
cat.	catalogue	illd.	illustrated
d.	died	illn.	illustration
diag./s.	diagram/s	Lib.	Library
drg./s.	drawing/s	n.d.	not dated
ed./s.	editor/s	pl.	plate, plates
edn./s.	edition/s	pub.	published, publisher's
endpprs.	endpapers	revd.	revised
engr.	engraved		

ACKNOWLEDGEMENTS

My thanks are due to the many people who have assisted, directly or indirectly, in the compilation of this bibliography, particularly the Librarians and staff of the following libraries: the British Library, Chartered Institute of Transport, Goldsmiths' Library (University of London), Institution of Civil Engineers, Leicester University, National Maritime Museum, and Science Museum Libraries. In addition, the following kindly allowed me unrestricted access to their private collections: the British Waterways Board, the Camping and Caravanning Club, the Cruising Association, Eric Garland, the Inland Waterways Association, Hugh McKnight, and Ronald Russell. Mike Miller generously lent me his own bibliographic notes.

I gratefully acknowledge the financial assistance towards the cost of part of the study provided by the Royal Society in the form of a Grant for Research in the History of Science.

1. General, including law and history

BEADON, George.
Ten minutes' reading of plain observations upon canals and navigable rivers; showing their vast importance . . .
London: Chapman & Hall, 1848. Frontis., [iv], 38, [1] pl.   B

BOARD OF TRADE. Canal Control Committee.
Handbook on canals: containing information relating to controlled canals, lists of the towns served by them, the names and addresses of public carriers.
London: HMSO, 1918. 27; fldg. map in pkt.   B
    Text of 1st edn. dated 1 January 1918. Revd. edn. dated 2 March 1918.
    2nd edn. November 1918. 1st edn. map shows England only. 2nd edn. map includes all Ireland but not Scotland.

BOOTH, Henry.
The carrying question stated, in reference to railways and canals . . .
Liverpool: Baines (printer), 1841. 32.   G

BOULTON, W[illiam] H[enry], b. 1869.
The pageant of transport through the ages.
London: Low et al., [1931]. xvii, 238, [65] pl.   B
      55-72 on inland waterways.

BOYLE, Edward, and WAGHORN, Thomas.
The law relating to traffic on railways and canals.
London: Clowes, 1901. 3 vols. xxxv, 449. vi. 463. xviii, 524.   B

BOYLE, Thomas.
Hope for the canals! Showing the evils of amalgamations with railways . . .
London: Simpkin et al., 1848. 43; illd.   B
    Includes proposals for 'Merchants and Manufacturers Carrying Union' and for use of Simpson's propellor.

BRADLEY, A[rthur] G[ranville], 1850-1943.
The rivers & streams of England.
London: Black, 1909. Frontis., xiii, 287, [4] (pub. cat.), 2-75 pl., fldg. map.   B
    Little on navigation.

[BRADSHAW, George.]
Appendix to G. Bradshaw's map of the canals and navigable rivers of the Midland Counties of England.
Manchester: Prentice (printer), 1829. 20.   B
    Lengths and falls.

[BRADSHAW, George.]
Lengths and levels to Bradshaw's maps of canals, navigable rivers, and railways.
London: White (printer), 1832. Fldg. map frontis., 15.   B
    South of England only.

[BRADSHAW, George.]
Lengths and levels to Bradshaw's maps of the canals, navigable rivers, and railways, in the principal part of England.
London: Ruff, 1833. 15, [5]-20.   B
    Most waterways north of Thames.
    Second part printed by Prentice.

BRIDGES, T[homas] C[harles].
Great canals.
London: Nelson, 1936. Frontis., vii, 104 (inc. pub. cat.), [5] pl.; map. (Discovery Books series).   B
    2nd edn. 1939.

BROOKE, H[enry], 1703(?)-1783.
The interests of Ireland considered . . . particularly with respect to inland navigation.
Dublin: Faulkner, 1759. 168, fldg. pl.   B

BURKITT, Robert.
Burkitt's observations on the inland navigation, draining of loughs, making entrances into ports . . .
Dublin: Cotter, 1755. [i], 8.   B

BUTTERWORTH, A[lexander] Kaye.
The practice of the Railway & Canal Commission, being the Railway & Canal Commission Rules,
1889, and the Railway and Canal Traffic Acts, 1854, 1873, and 1888.
London: Butterworth, 1889. v, 79, 32 (pub. cat.).   B

BUTTERWORTH, A[lexander] Kaye.
A treatise on the law relating to rates and traffic on railways and canals . . .
London: Butterworth, 1889. xxxii, 235, 108, 32 (pub. cat.).   B
    2nd edn. 1889.

CADBURY, George, and DOBBS, S[ealey] P[atrick].
Canals and inland waterways.
London: Pitman, 1929. xv, 160, 32 (pub. cat.); illd., map. (Pitman's Transport Library).   B

CANAL JOINT COMMITTEE.
Canals & inland waterways of England & Wales with carriers thereon. With details of the areas of
the six Regional Canal Committees.
London: CJC [c. 1945]. Map in pkt., 47, [6] maps (4 fldg.).   P
    Cover title 'Handbook on inland waterways'. Various issues.

CANAL JOINT COMMITTEE.
Post-war policy for the inland waterways . . .
London: CJC, 1945. 8.   T

CARY, John, c. 1754–1835.
Inland navigation: or select plans of the several navigable canals, throughout Great Britain: accom-
panied with abstracts of the different Acts of Parliament relative to them . . .
London: for the author, 1795. [iii], 132, [16] fldg. maps.   B

CAWLEY, George.
The future of British canals.
London: Wightman, 1902. 12.   C
    Reprinted from *The Manchester Guardian*.

CLIFFORD, Frederick, 1828–1904.
A history of Private Bill legislation.
London: Butterworth, 1885-7. 2 vols. xxviii, 507, 32 (pub. cat.); xvii, 966.   B
    Vol. I re-issued in 1887.
    Vol. I, 33–45 on canals; 467–476 on river navigations.

CONDER, Francis R.
Report on the comparative cost of transport by railway and canal.
London: Spottiswoode (printer), 1882. 16.   C

COULSON, H[enry] J[ohn] W[astell] and FORBES, Urquhart A[twell].
The Law relating to waters, sea, tidal and inland.
London: Sweet, 1880. xxvi, 748.   B
    6th edn. 1952.

CUNDY, N[icholas] W[ilcox]. b. 1778.
Inland transit. The practicability, utility, and benefit of railroads; the comparative attraction and
speed of steam engines, on a railroad, navigation, and turnpike road . . .
London: Hebert *et al.*, 1833. v, 161, [2] fldg. pl.   B
    2nd edn. 1834.

DARLINGTON, H[ayward] R[adcliffe].
The Railway and Canal Traffic Acts 1854 to 1888 . . .
London: Reeves & Turner: 1889. xl, 486.   B

DE MARÉ, Eric [Samuel], b. 1910.
The canals of England.
London: Architectural Press, 1950. 124; illd., map.   B
    4th (revd.) imp. 1965.

DE SALIS, Henry Rodolph.
Bradshaw's canals and navigable rivers of England and Wales: a handbook of inland navigation for
manufacturers, merchants, traders and others . . .
London: Blacklock, 1904. v–ix, 480, [8] (ads.); fldg. map in pkt.   B
   2nd edn. 1918. 3rd edn. 1928 (2 issues).
   Facs. of 1st edn. by David & Charles (Newton Abbot) 1969 (no map).

DE SALIS, Henry Rodolph.
A chronology of inland navigation in Great Britain . . . with an appendix giving the mileage travelled
by the author over the inland navigations of England.
London: Spon, 1897. [vi] , 133, 32 (pub. cat.).   B

DODD, Ralph.
A short historical account of the greater part of the principal canals in the known world; with some
reflections on the general utility of canals . . .
Newcastle-upon-Tyne: Charnley *et al.*, 1795. 27.   (B),N
   See also 'The engineering plagiarist'.

DUNWOODY, R[obert] B[rowne] , b. 1879.
The economic requirements for inland navigation transport in the British Isles.
London: Inst. Civ. Eng., 1920. iv, 84.   T
   Text of Vernon-Harcourt Lectures, 1920.

DUNWOODY, Robert Browne, b. 1879.
Inland water transport . . .
London: Assn. of Chambers of Commerce, 1913. 63.   N
   Text of paper to British Assn., 1913.

DUPIN, [Francois Pierre Charles] , 1784–1873.
The commercial power of Great Britain . . . view of the public works . . . streets, roads, canals,
aqueducts, bridges, coasts, and maritime ports.
London: Knight, 1825. 2 vols. xx, xlvi, [i] , |393; xvi, 399.   B
   Translated from French edn. of 1820–24.
   I: 89–343 on inland waterways.
      Accompanying atlas contains 10 pl., of which 4 show canal details.

EDWARDS, Lewis A[rthur] .
Inland waterways of Great Britain and Northern Ireland.
London: Imray *et al.*, 1950. Frontis., xvi, 440, pl., loose fldg. map.   P,(B)
   2nd edn. (1962) covers whole of Ireland.
   3rd edn. (1972), which omits whole of Ireland, is described as '5th edn.', and lists the 1950 and
   1962 edns. as 3rd and 4th, thus belatedly emphasising descent from Wilson's book of similar
   title (q.v.).

EGREMONT, John.
The Law relating to highways, turnpike-roads, public roads and navigable rivers; . . .
London: Longman *et al.*, 1830. [viii] , 419.   B
   This is Vol. I, but apparently no more published.
      196–208 on navigable rivers.

The engineering plagiarist: or, Dodd from Phillips exposed.
Newcastle: Whitfield (printer), [1795] . 21.   N

EYRE, Frank, and HADFIELD, [Ellis] Charles [Raymond] .
English rivers and canals.
London: Collins, 1945. 48, [8] pl.; illd., map. (Britain in Pictures Series).   B
   2nd imp. 1947.

[GRAHAME, Thomas.]
Correspondence between the Board of Trade and T. Grahame, Esq., late Chairman of the Grand
Junction Canal Company, on railway & canal combination.
London: Ridgway, 1852. 43.   B

GRAHAME, Thomas.
Essays and letters on . . . inland communication and transport.
London: Vacher & Son, 1835. Frontis., 61.   B
    Proposes that light, fast boats (as used on Scottish canals) should be more widely employed,
    and that lifts be used to bypass locks.

GRAHAME, Thomas.
A treatise on internal intercourse and communication . . . particularly in Great Britain.
London: Longman *et al.*, 1834, xiv, 160.   B
    160: 'End of the first part'.
    2nd part probably not published.

GYE, Percy, and WAGHORN, Thomas.
The new law of rates and charges on railways and canals, under the Railway & Canal Traffic Act 1888.
    London: Waterlow, 1889. vi, 101.   B

HADFIELD, [Ellis] Charles [Raymond], b. 1909.
British canals: an illustrated history.
London: Phoenix House, 1950. 259, VIII pl., illd., maps.   B
    RU edn. 1952. 6th edn. 1979.
    4th and earlier edns. exclude Ireland.

HAY'S WHARF CARTAGE CO.
Transport saga 1646–1947.
London: HWCC, 1947. Map, 64, map; illd.   P,N
    14–21 on Pickfords as canal carriers.

HEAD, George, 1782–1855.
A home tour through the manufacturing districts of England in the summer of 1835.
London: Murray, 1836 (new edn.). xi, 440, [2] (pub. cat.).   B
    Original edn. not located. New edn. (2 vol.) 1840. Facs. edn. by Cass (London) 1968.
        Describes river and canal trade, and several passenger boat journeys in Midlands, Scotland, etc.

HERCULES.
British railways and canals in relation to British trade and Government control.
London: Field & Tuer, [*c.* 1885]. [iii], 234, [2] (pub. cat.).   B

HICKMAN, Alfred.
A paper on railway rates & canal dues.
Wolverhampton: Barford & Newitt (printers), 1884. 8.   B

The history of inland navigations. Particularly those of the Duke of Bridgewater . . .
London: Lowndes, 1766. 2 parts. [vi], 88, [2] fldg. maps; [iv], 104.   B
    2nd edn. 'with additions' 1769.   3rd edn. 'with additions' 1779.
        Attributed to James Brindley.

HORNIMAN, Roy.
How to make the railways pay for the War; or the transport problem solved.
London: Routledge, 1916. Frontis., xx, 348, [3] pl., [2] fldg. pl.; diags.   B
    3rd edn. 1919.
        227–247 argue for better treatment of canals.

Industrial rivers of the United Kingdom.
See JONES, Evan Rowland.

INLAND WATERWAYS ASSOCIATION.
The future of the waterways.
London: IWA, 1947. 16; illd.   T

JACKMAN, W[illiam] T., b. 1871.
The development of transportation in modern England.
Cambridge: University Press, 1916. 2 vols. xvi, 459, [1] pl., [2] text maps, fldg. map in rear pkt.;
vii, 461–820, fldg. map in rear pkt.   B
    2nd (facs.) edn. by Cass (London) 1962.   3rd (facs.) edn. by Cass (London) 1966.

JEANS, J[ames] Stephen, 1846–1913.
Waterways and water transport in different countries: with a description of the Panama, Suez, Manchester, Nicaraguan, and other canals.
London: Spon, 1890. Frontis., xx, 507, [3] pl.; illd., maps, diags.   B

[JONES, Evan Rowland, (ed.), 1840–1920].
Industrial rivers of the United Kingdom; namely: the Thames, Mersey, Tyne, Tawe, Clyde, Wear, Taff, Avon, Southampton Water, the Hartlepools, Humber; Neath, Port Talbot, and Caermarthen; the Liffey, Usk, Tees, Severn, Wyre, and Lagan. By various well-known experts. London: T. Fisher Unwin, 1888. Frontis., x, 306, [6] (pub. cat.), [20] pl.   B
    Collection of articles first published in *The Shipping World*.

KANE, Robert [John], 1809–1890.
The industrial resources of Ireland.
Dublin: Hodges & Smith, 1844. xii, 417.   B
    2nd edn. 1845.
        Chap. 9 on transport.

LEACH, Edmund.
A treatise of universal inland navigations . . . demonstrating the possibility of making any river . . . navigable . . . without locks . . . Together with . . . a . . . machine, for inland navigation . . . which will raise, on an inclined plane, boats or lighters . . .
London: for the author, 1790. [v], vi, 201, 5 pl., extra pages (some fldg.) *58, **58, *61–*69, *198.   B
    Another issue by Hamilton (London) 1791 recorded in N.

M[ACLAREN], C[harles], 1782–1866.
Railways compared with canals & common roads, and their uses and advantages explained: . . .
Edinburgh: Constable, 1825. 66.   B
    Reprinted in *The Pamphleteer*, Vol. XXVI, 1826.

MANSION HOUSE ASSN. ON RAILWAY & CANAL TRAFFIC.
Canals and inland navigations. Report of a meeting convened by the . . .
London: Singer, 1905. 49; fldg. table.   B

MINISTRY OF INFORMATION (for Ministry of War Transport).
Transport goes to war: the official story of British transport, 1939–1942.
London: HMSO, 1942. 80; illd.   B
    38–43 on inland waterways.

MORRIS, E. A. Montmorency.
Canals and waterways of western Europe.
Dublin: Sealy *et al.*, 1905. 18, [2] maps.   B
    Text of paper read to Statistical & Social Inquiry Soc. of Ireland. Largely devoted to British Isles.

NATIONAL COUNCIL FOR INLAND WATERWAYS.
Official handbook.
Birmingham: NCIW, 1926. 64 (inc. ads.); illd.   C
    Cover title 'Canals and inland waterways'.

NEWTON, H.
British canals.
London: Allan, 1948. 64, fldg. map; illd.   B
    Includes Grand Union Canal Co. fleet list.

O'BRIEN, W.
The prize essay on canals and canal conveyance.
London: Weale, 1858. vii, 36, III fldg. pl.   B
    This won the Canal Association prize for 1856, the first year it was offered.

Our canals. Reprint of special article in *The Times* of 16 May 1894.
[?]: [?], [1894?]. 8.   P

PALMER, J[oseph] E[dward], 1856–1910.
British canals: problems and possibilities.
London: Unwin, 1910. 254.   B

PALMER, J[oseph?] E[dward?].
How to cheapen inland transit.
Dublin: Hodges *et al.*, 1896. 12.   L

PHILLIPS, J[ohn].
A general history of inland navigation, foreign and domestic: containing a complete account of the canals already executed in England . . .
London: Taylor, 1792. [ii], xx, 373, 4 pl., fldg. map.   N
   First 3 edns. 4to.; 2nd (1793) and 3rd (1795) formed by adding supplements at rear. 4th (1803) and 5th (1809) edns. 8vo. Various issues of both latter.
      Facs. edn. of 5th by David & Charles (Newton Abbot) 1970.

PRATT, Edwin A., 1854–1922.
British canals: is their resuscitation practicable?
   London: Murray, 1906. Frontis., xi, 159 [2] (pub. cat.), [9] pl., [3] plans.   B

PRATT, Edwin A., 1854–1922.
Canals and traders: the argument pictorial as applied to the report of the Royal Commission on Canals and Waterways.
London: King, 1910. Frontis. fldg. map, xi, 123, [1] (pub. cat.), [34] pl., [2] fldg. pl; text diags.   B

PRATT, Edwin A., 1854–1922.
A history of inland transport and communication in England.
London: Paul *et al.*, 1912. xii, 532.   B
   Facs. edn. by David & Charles (Newton Abbot) 1970.

PRATT, Edwin A., 1854–1922.
Railways and their rates: with an appendix on the British canal problem.
London: Murray, 1905, ix, 361, [5] pl.   B
   2nd edn. 1906.

PRATT, Edwin A., 1854–1922.
Scottish canals and waterways.
London: Selwyn & Blount, 1922. xi, 299, [4] (pub. cat.), [2] fldg. pl.; maps, diags.   B

PRIESTLEY, Joseph, 1766 or 67–1852.
Historical account of the navigable rivers, canals and railways, throughout Great Britain, . . .
London: Longman *et al.*, 1831. Frontis. map, [iv], xii, 777, x, fldg. pl.   B
   This is 4to; 2nd edn. (1831) 8vo.
      Facs. of 1st edn. by Cass (London) 1967, described as '2nd edn.'. Facs. of 2nd edn. by David & Charles (Newton Abbot) 1969.

PYNE, W. H.
Microcosm: or, a picturesque delineation of the arts, agriculture, manufacture, &c. of Great Britain . . .
London: Pyne, 1803. Vol. I. [iv], [90] pl.   B
   2nd Vol. (1806) not seen. 2nd edn. 1806–8.
      Several drgs. of barges, locks.

RODGERS, John.
English rivers.
London: Batsford, 1947/8. Frontis., viii, 184, [99] pl.; map endpprs.   B

ROLT, L[ionel] T[homas] C[aswall], 1910–1974.
Inland waterways.
London: Assn. for Planning and Regional Reconstruction, 1946. 12. (Report R41).   P

ROLT, L[ionel] T[homas] C[aswall], 1910–1974.
The inland waterways of England.
London: Allen & Unwin, 1950. Frontis., 221, XLVIII pl., fldg. map.   B
   5th imp. 1970. 3rd (1962) and later imps. have new preface.
   2nd edn. 1979.

SALT, Samuel.
Statistics and calculations essentially necessary to persons connected with railways or canals . . .
Manchester: Bradshaw & Blacklock, 1845. 116.   B

SANER, J[ohn] A[rthur].
Digest of the fourth and final report of the Royal Commission on canals and waterways . . .
Northwich: Weaver Navn. Trustees, 1910. [i], 36.   C

SKRINE, Henry, 1755–1803.
A general account of all the rivers of note in Great Britain . . .
London: Elmsly, 1801. Double frontis., xx, 412, [17] maps.   B

SOUTH, Frederick Charles.
The British and Irish waterways gazetteer. Comprising a directory of steamship, hoy and canal companies, together with a list of London wharves and licensed lightermen; also about seven thousand water routes to over two thousand important places in the United Kingdom.
London: Stanmer, 1910. xiv (inc. ads.), 381.   P

SPENCE, Peter
How the railway companies are crippling British industry and destroying the canals: with suggestions for . . . rescuing the water ways permanently for the nation. Being the evidence given . . . before the House of Commons' Committee on Railway Rates (1881–2), by Peter Spence . . .
Manchester: Tubbs et al., [1882?]. 28.   N,P

SUTCLIFFE, John.
A treatise on canals and reservoirs . . . with observations on the Rochdale, Leeds and Liverpool, and Huddersfield Canals . . . and also on the Bridgewater, the Lancaster, and the Kennett and Avon Canals . . .
Rochdale: for the author, 1816. [v], xiv, 414.   B
    73–376 on canals.

TATHAM, W[illiam], 1752–1819.
The political economy of inland navigation, irrigation and drainage . . .
London: Faulder, 1799. xvi, 500, [3] (app.), VI pl., [5] fldg. plans.   B

TATHAM, William, 1752–1819.
Remarks on inland canals, the small system of interior navigation, various uses of the inclined plane, &c. &c. in a letter from William Tatham to a proprietor in the Colebrook-Dale and Stratford Canals.
London: Taylor, 1798. 20, [1] pl.   B

THOMPSON, H[ubert] Gordon.
The canal system of England: its growth and present condition, with particular reference to the cheap carriage of goods.
London: Unwin for Cobden Club, 1902. [iii], 70, iv (index).   B
    2nd edn. 1904.

A treatise on inland navigation.
Salisbury: Collins (printer), 1788. 44.   B

WEBSTER, Robert G[rant], 1845–1925.
The Law relating to canals: comprising a treatise on navigable rivers and canals . . .
London: Stevens, 1885. Fldg. map frontis., lxxiii, 361, 32 (pub. cat.).   B

WELLS, Lionel R.
A sketch of the history of the canal and river navigations of England and Wales and of their present condition, with suggestions for their future development.
Manchester: [?], 1894. 187–204, fldg. map.   C
    Offprint from Mem. Proc. Manchester Lit. Phil. Soc.

WILLAN, T[homas] S[tuart].
River navigation in England 1600–1750.
London: Oxford Univ. Press, 1936. [viii], 163; V maps. (Oxford Historical Series).   B
    Facs. edn. by Cass (London) 1964.

WILLIAMS, C[harles] W[ye], 1779–1866.
Observations on an important feature in the State of Ireland . . . with a description of the navigation of the River Shannon.
London: Vacher, 1831. iv, iv, fldg. map, 59.   B
   See next entry for 2nd edn.

WILLIAMS, C[harles] W[ye], 1779–1866.
Observations on the inland navigation of Ireland . . . with a description of the River Shannon . . .
London: Vacher, 1833, (2nd edn.). viii, 108, fldg. plan.   B
   See previous entry for 1st edn.

WILLIAMS, Ernest Edwin George, 1866–1935.
A manual of the law and practice of the Railway and Canal Commissioners' Court . . .
London: Butterworth, 1913. xxi, 292, 14 (index).   B

WILSON, W[illiam] Eric.
Inland waterways of Great Britain. A handbook . . . of the principal canals and rivers of England and Scotland . . .
London: Imray *et al*. 1939. [viii], 179, loose fldg. map.   B
   2nd edn. 1947.

WOODFALL, Robert.
The new law and practice of railway and canal traffic. Being the Railway and Canal Traffic Act, 1888 . . .
London: Clowes, 1889. xvi, 221.   B

WOOLRYCH, Humphrey W[illiam], 1795–1871.
A treatise on the Law of waters . . . including . . . rivers, canals, dock companies . . .
London: Saunders & Benning, 1830. xxxi, 501, [2] (pub. cat.)   B
   2nd edn. 1851.

WORDSWORTH, Charles [Favell Forth], 1803–1874.
The Law of railway, canal, water, dock, gas, and other companies, requiring express authority of Parliament . . .
London: Benning, 1851 (6th edn.) xxvi, 533, 623 (App., index).   B
   Earlier edns. exclude canals.

2: *Histories and general accounts of particular areas or waterways*

AIKIN, J[ohn].
A description of the country from thirty to forty miles round Manchester.
London: Stockdale, 1795. Frontis., extra engr. title, xxiv, 624, [63] maps and pl. (2 fldg.).   B
   Facs. edn. by David & Charles (Newton Abbot) 1968.
      105–45, 582–3 on inland waterways.

AIRE & CALDER NAVIGATION [CO.]
Aire & Calder Navigation: the water route between Lancashire & Yorkshire and the East Coast.
Leeds: ACN, [*c*. 1923]. [i], 30, fldg. pl.; illd., map.   T
   Various issues.

ALLCROFT, A[rthur] Hadrian, 1865–1929.
Waters of Arun.
London: Methuen, 1930. Frontis., xv, 170, [5] pl., 2 fldg. maps; 19 text maps.   B

ALLEN, [Edgar] Fletcher, b. 1886.
The fair rivers of southern England.
London: Muller, 1943. 11, [64] pl.   B
   3rd edn. [imp?] 1946. Photos had all previously appeared in AUSTIN (q.v.).
      Roughly south of the basins of the R. Trent and R. Mersey; includes Wales.

ALLNUTT, Z[achariah].
Useful and correct accounts of the navigation of the river and canals west of London . . . second
edition, much improved.
Henley: for the author, [c. 1810?]. 20, fldg. map.   B
   1st edn. not seen.

ARMSTRONG, Walter, 1850-1918.
The Thames from its source to the sea.
London: Virtue, [1886-7]. 2 vols. Frontis., engr. title, iv, 176, [6] pl., map; illd. Frontis., engr. title,
iv, 176, [5] pl., map; illd.   B
   Some issues of Vol. I, and all issues of Vol. II, are titled 'The Thames from its rise to the Nore'.

AUSTIN, A. B.
Rivers of the south.
London: Muller, 1938. 200, [64] pl.; map endpprs.   B
   All plates reprinted in Fletcher ALLEN's 'The fair rivers of southern England' (q.v.).

BADESLADE, Thomas.
The history of the ancient and present state of the navigation of the port of King's-Lyn, and of
Cambridge, . . . and of the navigable rivers that have their course through the great-level of the Fens,
called Bedford Level . . .
London: for the author, 1725. [xii], 148, [7] pl, some fldg.   B
   2nd edn. 1766.

BAKER, B[ernard] Granville.
Blithe waters: sheaves out of Suffolk.
London: Granton, 1931. 259; illd., map endpprs.   B
     General account of Suffolk rivers.

BAKER, B[ernard] Granville.
Waveney.
London: Allan, 1924. Frontis., [iv], 254; illd.   B
     Little on navigation.

BARKER, T. C.
The Sankey Navigation: the first Lancashire canal.
[?]: [?], [1948]. 121-55, [7] pl.; [2] maps.   L
     Reprinted from *Trans. Hist. Soc. Lancs. Ches.*

BARRON, James.
A history of the Ribble Navigation from Preston to the sea.
[Preston]: for the Corporation, 1938. Frontis., xv, 503; illd., maps.   B

BARROW, Walter, and BARROW, Richard.
Fladbury and its mills and the Lower Avon Navigation.
Birmingham: Kynoch Press (privately printed), 1949. 32; illd.   P

BELLOC, [Joseph] Hilaire [Pierre], 1870-1953.
The historic Thames.
London: Dent, 1907. Frontis., vii, 223, [58] pl.   B
   Several edns.

BLOWER, Benjamin.
The Mersey ancient and modern.
Liverpool: Howell, 1878. Frontis., vii, 88, [22] pl., [3] fldg. maps.   B

BOYDELL'S Thames. *See* COMBE, William.

BRADLEY, A[rthur] G[ranville], 1850-1943.
The Avon and Shakespeare's country.
London: Methuen, 1910. ix, 365, 31 (pub. cat.), [30] pl.   B
   2nd edn. 1913.
     Little on navigation.

144   CANALS – A NEW LOOK

BRADLEY, A[rthur] G[ranville], 1850-1943.
A book of the Severn.
London: Methuen, 1920. Frontis., viii, 351, 31 (pub. cat.), [15] pl.   B
    Little on navigation.

BRADLEY, A[rthur] G[ranville], 1850-1943.
The Wye.
London: Black, 1910. Frontis, viii, 189, [2] (pub, cat.), [23] pl., fldg. map.   B
    2nd edn., 1926, with 16 pl.
      Abridged edn. 1916.
        Little on navigation.

BRADLEY, Tom.
The Aire.
Leeds: Yorkshire Post, 1893. 2 pts. Fldg. pl., 38, [4] (pub. ads.). Fldg. pl., 53, [5] (pub. ads.).
(Yorkshire Rivers Series No. 9).   B
    Reprinted from *Yorkshire Weekly Post.*

BRADLEY, Tom.
The Derwent.
Leeds: Yorkshire Post, 1891. Fldg. pl., 50, [2] (pub. ads.). (Yorkshire Rivers Series No. 6).   P,(B)
    Reprinted from *Yorkshire Weekly Post.*

BRADLEY, Tom.
The Ouse.
Leeds: Yorkshire Post, 1891. Fldg. pl., 46, [2] (pub. ads.). (Yorkshire Rivers Series No. 5).   P,(B)
    Reprinted from *Yorkshire Weekly Post.*

BRADLEY, Tom.
The Swale.
Leeds: Yorkshire Post, n.d. (*c.* 1891). Fldg. pl. 36, [4] (pub. ads.). (Yorkshire River Series
No. 4).   P,(B)
    Reprinted from *Yorkshire Weekly Post.*

BRADLEY, Tom.
The Ure.
Leeds: Yorkshire Post, 1891. (2nd edn.). Fldg. pl., 46, [2] (pub. ads.). (Yorkshire Rivers Series
No. 3).   P,(B)
    Reprinted from *Yorkshire Weekly Post.*

BRADLEY, Tom.
The Wharfe.
Leeds: Yorkshire Post, [*c.* 1890] 3rd edn. Fldg. pl., 43, [5] (ads.). (Yorkshire River Series
No. 1).   B
    Reprinted from *Yorkshire Weekly Post.*

BROWN, James.
The present state and future prospects of the Monmouthshire Canal Company considered; in a
letter addressed to the Committee of Management.
London: Weale, 1847. 37, [2] pl.   B
    Relates entirely to the Newport and Pontypool rly.

BURT, Edward.
Letters from a gentleman in the north of Scotland to his friend in London . . .
London: Birt, 1754. 2 vols. Plan frontis., x, 344. [i] , 368.   B
    5th edn. 1818.
      Vol. 2, 330-336 re Govt. vessel on L. Ness, and proposed Caledonian Canal.

CADELL, H[enry] M[oubray].
The story of the Forth.
Glasgow: Maclehose, 1913. Frontis., xvii, 299, [45] pl., [9] maps; drgs.   B
    Little on navigation.

CALDER & HEBBLE NAVIGATION.
Calder & Hebble Navigation: general information . . .
Halifax: CHN, [c. 1937]. 56; illd., maps.   T

CALVER, E[dward] K[illwick].
River Tyne. A letter to the Tyne Improvement Commissioners, . . . with replies . . .
Newcastle-upon-Tyne: Journal Office, 1852. 16, 19, 8.   B

COLOQUHOUN, P[atrick], 1745-1820.
A treatise on the commerce and police of the River Thames . . . and suggesting . . . a legislative
system of River Police.
London: Mawman, 1800. xxxiv, 676, [20] (index), [2] fldg. tables.   B

COMBE, William, 1742-1823.
An history of the principal rivers of Great Britain.
London: Boydell, 1794-6. 2 vols. Frontis., xvii, 312, 46 pl. 2 fldg. maps. [viii], 294, 30 pl.   B
   Each vol. has an additional title page reading 'An history of the River Thames'. No further
   volumes published. Later issues lack this additional title.

A companion for canal passages, betwixt Edinburgh and Glasgow; giving a complete account of . . .
objects . . . seen along the line of the two canals . . .
Edinburgh: Aitken, 1823. 36.
   Facs. edn. by Linlithgow Union Canal Soc., 1981, from original (not seen) in Mitchell Lib.,
   Glasgow.

COOK, C[harles] H[enry], 1858-1933.
Thames rights and Thames wrongs: a disclosure, with notes explanatory and critical on the Thames
Bill of 1894.
London: Constable, 1894. xvi, 227, [2] (ads.), 16 (pub. cat.).   B
   2nd edn. 1895.

COOKE, William Bernard, 1778-1855.
The Thames . . .
London: Vernor et al., 1811. 2 vols. [137], [46] pl.; [172], [38] pl.   B

COOKE, W[illiam B[ernard], 1778-1855.
Views on the Thames.
London: Cooke, [1812-18]. 2 vols. Frontis., [374]. [75] pl. (no text).   B
   1822 edn. combines plates and text in 1 vol.

CORBETT, J.
The River Irwell: pleasant reminiscences of the nineteenth century and suggestions for improvements
in the twentieth.
Manchester: Heywood, 1907. [iii], 155, xii (index etc.), [47] pl., [3] maps (1 fldg.).   B
   Chap. 4 on Bridgewater, and Chap. 7 on Manchester, Bolton & Bury Canal.

DE SALIS, Henry Rodolph.
Norfolk waterways: a guide to the navigable waterways of the Norfolk Broad district, compiled
after a personal survey of the whole.
London: Jarrold, 1900, 65, [3] (ads.).   B

DUNSTON, G.
The rivers of Axholme, with a history of the navigable rivers and canals of the district.
London: Brown [1909]. Frontis., xiii, 155, [5] pl. (4 fldg.).   B

DUTT, William A[lfred].
The Norfolk Broads.
London: Methuen, 1903. Frontis., viii, 379, [73] pl., fldg. map; 3 figs.   B
   3rd edn. 1930.

EVANS, J. T.
Notes on the history of the navigation of the River Trent.
Nottingham: Trent Navn. Co., 1945. 25; map.   P

FALK, Bernard.
The Bridgewater millions: a candid family history.
London: Hutchinson, 1942. Frontis., 247, [9] (pub. cat.), [32] pl.   B

FARRAR, C[harles] F[rederick], b. 1860.
Ouse's silent tide.
Bedford: Arts Club, 1910. 23; map.   B
    History of Gt. Ouse.

FEARNSIDE, William Gray.
Eighty picturesque views on the Thames and Medway, engraved on steel by the first artists . . .
London: Black & Armstrong, [c. 1834]. Engr. title, engr. dedication, [4], iv, 84, [79] pl., fldg.
map.   B
  B copy defective. Pub. in monthly parts. Engr. title is 'Tombleson's Thames'.
  Facs. edn. by Bishopsgate Press (London) c. 1980.

FENN, E[ric] A[lfred] Humphrey.
The origin of the inland waterways of East Anglia.
London: Blake's, [c. 1934]. 16.   B

FINCH, William Coles.
The Medway, river and valley.
London: Daniel, 1929. Frontis., 239, [3] (pub. cat.), [76] pl; drgs.   B

FOSTER, A[lbert] J[ohn], b. 1844.
The Ouse.
London: Soc. for Promoting Christian Knowledge, [c. 1891]. xv (inc. frontis.), fldg. map, 223, 8
(pub. cat.); illd.   (B),P
    Great Ouse.

FOWLER, Joseph (ed.).
A description of the high stream of Arundel . . . written . . . about . . . 1637 . . .
Littlehamptom: Nature & Archaeology Circle, 1929. 71, fldg. map in pkt.   B
    Little on navigation.

FRENCH, William.
The Port of Glasgow.
Glasgow: Clyde Navigation Trust, 1947. 218, 291 (ads.); illd.   P

GILPIN, William, 1724-1804.
Observations on the River Wye, and several parts of South Wales . . . [in] 1770.
London: Blamire, 1782. [iii], xi, 99, [15] pl.   B
    5th edn. 1800.
      6-43 boat trip downstream from Ross.

GLOUCESTER & BERKELEY CANAL CO.
The Gloucester & Berkeley Canal.
Gloucester: GBC, 1826. Map frontis., 13.   B

GRAHAME, Thomas.
Letters to the Directors of the canal navigation connecting London with Northamptonshire,
Leicestershire, and other Midland counties, on the subject of the traffic in live stock, horses,
carriages &c., &c.
[?]: privately printed, 1854. 24, fldg. map.   B

GRAND UNION CANAL CO.
Making transport history: Grand Union Canal Company's new all water routes.
London: GUCC, [c. 1935]. 120 (inc. ads.); illd., maps.   P

GRAND UNION CANAL CO.
The waterways of England.
London: GUCC, [c. 1935]. 121 (inc. ads.); illd., maps.   P
  Cover title 'Arteries of commerce'.

GUTHRIE, James, 1826[?]–1880.
The River Tyne: its history and resources.
Newcastle: Reid, 1880. [i], frontis., xvi, 248, [7] (pub. cat.), [7] pl., [7] fldg. pl.   B

HALL, Mr. & Mrs. S[amuel] C[arter].
The book of South Wales, the Wye, and the coast.
London: Hall *et al.*, 1861. xi, 512, [4] (pub. cat.); illd.   B
   Facs. edns. by James (Teddington) 1979, 1980.
     14-19 on R. Severn and Berkeley Ship Canal.

HALL, Mr. & Mrs. S[amuel] C[arter].
The book of the Thames, from its rise to its fall.
London: Hall *et al.*, 1859. xii, 516; illd.   B
   Reprinted from the *Art Journal*.
   Several issues including one with mounted photographs pub. by Cassell (London) 1869.
   2nd edn. by Virtue (London) 1877.
   5 facs. edns. by James (Teddington) 1975-1978.

HARRAL, Thomas, d. 1853.
Picturesque views of the Severn: with historical and topographical illustrations.
London: Whittaker, 1824. 2 vols. Frontis., viii, 306, [28] pl.; [iii], frontis., 290, [21] pl.   B

HASSELL, J[ohn], d. 1825.
Tour of the Grand Junction, illustrated in a series of engravings . . .
London: for the author, 1819. Frontis., viii, 147, [4] (index), [23] pl.
   Although generally coloured, copies exist with the plates uncoloured. An edited edn. by
   Cranfield & Bonfiel (London) 1961.

HEYWOOD, John.
John Heywood's illustrated Manchester Ship Canal route guide.
Manchester: Heywood, [*c*. 1893]. [vi] (ads.), 32, fldg. map, [vi] ads; illd.   B

HEYWOOD, John.
John Heywood's visitors' illustrated guide to the Ship Canal, Eastham, Pool Hall, and Ellesmere Port . . .
Manchester: Heywood, [*c*. 1890]. [xi], 33, [viii] (ads.); illd.   B
   Various edns., one of 1892 having 32-34 on Weaver.

HOBSON, John Morrison.
The book of the Wandle: the story of a Surrey river.
London: Routledge, 1924. xii, 196, XVI pl.; illd.   B
    Little on navigation.

HOWSON, J[ohn] S[aul], 1816-1885.
The River Dee: its aspect and history.
London: Virtue *et al.*, 1875. xiv, 174; illd.   B
   This is 4to; 2nd (1887) and 3rd (1889) edns. 8vo.

Illustrated history of the Manchester Ship Canal during construction and complete.
Manchester: Heywood, [1894]. 16; illd.   B

INSH, George Pratt.
The elusive river: a roving survey of the Clyde from Daerhead to the Tail of the Bank.
Edinburgh: Moray Press, 1933. [i], frontis., 198, [2] (pub. cat.), [3] pl.; map endpprs.   B
   3rd edn. 1946. Later edns. have slightly different titles.

IRELAND, Samuel, d. 1800.
Picturesque views on the River Medway . . .
London: Egerton, 1793. Engr. title, xiii, map, 206, [28] pl.; drgs.   B

IRELAND, Samuel, d. 1800.
Picturesque views on the River Thames . . .
London: Egerton, 1792. 2 vols. Engr. title, xvi, map, 209, [1] (errata), [27] pl.; drgs. Engr. title, vii,
map, 258, [1] (errata), [1] (ad.), [25] pl.; drgs.   B
   Another edn. 1801-2.

IRELAND, Samuel, d. 1800.
Picturesque views on the River Wye . . .
London: Faulder & Egerton, 1797. xii, map, 159, [31] pl.   B

IRELAND, Samuel, d. 1800.
Picturesque views on the Upper, or Warwickshire Avon, from its source at Naseby to its junction with the Severn at Tewkesbury . . .
London: Faulder & Egerton, 1795. Frontis., xviii, 284, map, [31] pl.   B

JOHNSON, R[obert] W[illiam] , b. 1863.
The making of the Tyne: a record of fifty years' progress . . .
London: Scott, 1895. xii, [2] fldg. charts, 360; illd.   B

JOHNSON, R[obert] W[illiam] , & AUGHTON, Richard (eds.).
The River Tyne, its trade and facilities: an official handbook . . .
Newcastle: Reid, 1925. lix (ads.), 108, 2 fldg. plans; illd.   B
    Other edns. 1930, 1934.

JONES, L[lewellyn] Rodwell, 1881-1947.
The geography of London River.
London: Methuen, 1931. xi, 184, IV pl. (inc. frontis.); maps, diags.   B

KILLICK, H. F.
Notes on the early history of the Leeds and Liverpool Canal . . .
Bradford: Gaskarth (printer), [c. 1897] . Fldg. map, [i], 73.   P
    Reprinted from *Bradford Antiquary*.

[KINDERLEY, Nathaniel (senior).]
The present state of the navigation of the towns of Lyn, Wisbeech, Spalding, and Boston . . .
Bury St Edmunds, for the author, 1721. 20.   N
    2nd edn. pub. by Noon (London) 1751, edited by his son, Nathaniel Kinderley (junior).

LACY, Thomas J. (ed.).
The origin of the River Wey Navigation. Being an account of the canalization of the river, from a manuscript written by Richard Scotcher, in 1657. Now first published.
Guildford: Curtis, 1895. Frontis., [i], 27.   B

LEE CONSERVANCY BOARD.
General information respecting the navigation . . . and the River Lee.
London: LCB, 1934. Map, 8.   P

LEE CONSERVANCY BOARD.
General information respecting navigation . . . and traffic on the River Lee & River Stort.
London: LCB, 1946. 56; illd., maps, ads.   L

LEECH, Bosdin [Thomas] . b. 1836.
History of the Manchester Ship Canal . . .
Manchester: Sherratt & Hughes, 1907. 2 vols. Frontis, xv, 333, [26] pl., [9] fldg. pl., [5] fldg. plans in rear pkt.; illd. Frontis., ix, 351, [51] pl., [6] fldg. pl., [2] fldg. plans in rear pkt.; illd.   B

LEEDS & LIVERPOOL CANAL CO.
The waterways of England . . .
Liverpool: LLCC, [c. 1936] . 79; illd., maps.   P
    Cover title 'By waterway – the betterway 'twixt Liverpool and Leeds'.

LESLIE, George D[unlop] , b. 1835.
Our river.
London: Bradbury *et al.*, 1881. Frontis, xvi, 272, [12] pl.; illd.   B
    2nd edn. 1888.
        The Thames.

LINNEY, A[lbert] G[ravely] .
Lure and lore of London's river.
London: Low *et al.*, [1932]. Frontis., xxii, 242, [24] pl., [8] maps.   B
    163-191 on connecting waterways.

LLOYD, John (ed.).
Papers relating to the history and navigation of the Rivers Wye and Lug.
Hereford: Hereford Times, 1873. [iii], 48.   B

London's Canals.
[London?: ?. 1925?]. 24.   P
   Reprinted from the *Daily Telegraph* of 22 to 29 June 1925.
   Facs. edn. by Danetre Press (Daventry) [c. 1975].

MACARTHUR, [David] Wilson.
The River Fowey.
London: Cassell, 1948. Frontis., ix, 180, [32] pl., map.   B

MACGREGOR, John.
A letter to the merchants, coalowners, and ship-owners of Newcastle, on the present state of the
conservatorship of the Tyne.
Newcastle-upon-Tyne: Mitchell (printers), 1832. 21.   B

MACKAY, Charles, 1814-1889.
The Thames and its tributaries . . .
London: Bentley, 1840. 2 vols. Engr. title, xii, 400; illd. Engr. title, viii, 412; illd.   B

MANCHESTER EDUCATION COMMITTEE.
The inland port of Manchester; its ships & their cargoes.
Manchester: Port of M. Committee of the Chamber of Commerce, 1938. 37 (last leaf fldg.); illd.,
maps.   B

MANCHESTER SHIP CANAL CO.
Handbook, schedule of rates & charges, &c.
Manchester: MSC, 1894. [2] fldg. pl., 69.   B

Manchester Ship Canal: 12 photogravure views.
[Manchester?]: [Heywood?], [1891?]. [12] pl., no text.   B

Manchester Ship Canal: 12 'Photo Print' views.
[Manchester]: [Heywood?]. 12 pl., no text.   B

Manchester Ship Canal: 18 'Photo Print' views.
Manchester: Heywood, c. 1894. [18] pl., no text.   B

MAXWELL, Donald, b. 1877.
The book of the Clyde . . . from its source to the firth.
London: Lane, 1927. Frontis., ix, 179, [2] (pub. cat.), [1] pl.; drgs.   B

MAXWELL, Donald, b. 1877.
A pilgrimage of the Thames.
London: Centenary Press, 1932. xii, 193; illd., maps.   B

MAXWELL, Gordon S[tanley].
The fringe of London.
London: Palmer, 1925. Frontis., 254, [1] (pub. ad.); drgs.   B
    241-251 on Grosvenor Canal.

MURRAY, John Fisher, 1811-1865.
A picturesque tour of the River Thames in its western course; including particular descriptions of
Richmond, Windsor, and Hampton Court.
London: Bohn, 1849. [vii]-x, 356, [4] maps; illd.   B
   N records an edn. of 1845. Other edns.: 1853 (N), 1862 (B).

OLDHAM, Arthur Artis.
A history of Wisbech River.
Wisbech: Oldham, 1933. [ix], 114, [6] pl.   S

PALMER, William T.
The River Mersey.
London: Hale, 1944. Frontis., 250, [14] pl., fldg. map.   B
  2nd edn. 1950.
     115–123 on connecting canals.

PANKHURST, [Richard Marsden].
The Manchester Ship Canal. Opening for traffic. Speech of Dr. Pankhurst, on application to the
County Quarter Sessions ... 30th Dec., 1893, for a certificate declaring the Manchester Ship Canal
completed and fit for the reception of vessels.
Manchester: Heywood, 1894, 30.   B
    Also contains translations into French and German.

Panoramic views of the Manchester Ship Canal taken at the opening ceremony, January 1st,
1894.
[Manchester?]; [Heywood?], [1894]. [6] pl., no text.   B

PENNANT, Thomas, 1726–1798.
The journey from Chester to London.
London: White, 1782. Engr. title, iv, 458, XXII pl.   B
   Further edns.: White (London) 1783; Wilkie & Robinson (London) 1811.
     50–56 *re* Trent & Mersey Canal.

Photogravure views of the Manchester Ship Canal.
[Manchester?]: [Heywood?], [c. 1891]. Fldg. strip of [12] pl.   B

Picturesque tour of the River Thames; illustrated by twenty-four coloured views, a map and vignettes
... by William Westall and Samuel Owen.
London: Ackermann, 1828. vi, fldg. map, 172. [24] pl.   B

PLYMLEY, Joseph.
General view of the agriculture of Shropshire ...
London: Phillips, 1803. xxiv, 366, 2 (pub. cat.), [6] pl. (5 fldg.); map, drgs.   B
    284–316 and 4 pl. are by Telford and describe inland waterways.

PONTEY, William.
A short account of the Huddersfield Canal ... with some remarks applicable to the projected London
and Cambridge Canal.
Cambridge: Hodson, 1802. [i], 41.   C

The present state of the navigation of the towns of Lyn, Wisbeech ...
See KINDERLEY, Nathaniel (Senior).

PROTHERO, Thomas.
... letter to the Monmouthshire Railway and Canal Company ...
Newport: Christophers, 1851. 11.   B

QUILLER-COUCH, A[rthur] T[homas], 1863–1944.
The Warwickshire Avon.
London: Osgood, McIlvaine & Co., 1892. 144; illd.   B

Report to the general assembly of the Ellesmere Canal proprietors, ... to which is annexed, the
oration, delivered at Pontcysylte Aqueduct, on its first opening, November 26, 1805.
Shrewsbury: for the Co., 1806. Fldg. map frontis., 43, 36.   C

RILEY, Frederic.
The Ribble from its source to the sea.
Settle: Lambert, 1914. Frontis., xvi, 230, 20 pl. 4 fldg. maps.   B

RIVER SEVERN DEVELOPMENT ASSOCIATION.
Sites on the River Severn, England's longest river.
Birmingham: RSDA, [c. 1932]. 44; fldg. pl.; maps, illd.   B
  2nd edn. 1936.

The rivers of Great Britain — descriptive, historical, pictorial. Rivers of the East Coast.
London: Cassell, 1889. Etching frontis., viii, 376, 8 (pub. cat.); illd., maps.   B
   Subscribers edn. (n.d.) 2 vols., each with 6 etchings.
   Other edns. 1892, 1902. Minor variations in title between issues.

The rivers of Great Britain — descriptive, historical, pictorial. Rivers of the South and West Coasts.
London: Cassell, 1897. Frontis., viii, 376, [4] (pub. cat.); illd., maps.   B
   Another edn. 1901[?]. Minor variations in title between issues.

The rivers of Great Britain — descriptive, historical, pictorial. The Thames from source to sea.
See The royal river.

ROBBERDS, J[ohn] W[arden] (Junior).
Scenery of the rivers of Norfolk, comprising the Yare, the Waveney, and the Bure . . .
London: Moon *et al.*, 1834. unpaginated; illd.   B

ROBERTSON, H[enry] R[obert], 1839–1921.
Life on the upper Thames.
London: Virtue *et al.*, 1875. Frontis., x, 214; illd.   B

The royal river: the Thames, from source to sea. Descriptive, historical, pictorial.
London: Cassell, 1885. Frontis., viii, map, 368; illd., maps.   B
   Orig. issued in parts. Other edns., with variations in title, 1891, 1902.

SAMPSON, Charles.
Ghosts of the Broads.
London: Yachtsman Pubg. Co., 1931. 195; drgs., maps.   B
   New edn. by Jarrold (Norwich) 1973.

SANER, J[ohn] A[rthur], b.1864.
A short description of the Weaver Navigation.
[Northwich]: [WN Trustees], 1914 (4th edn.). 16; illd.   P
   No other edn. seen. 1st edn. 1902.

SAXTON, A. H. (pub.).
Birmingham road and canal transport guide.
Birmingham: Saxton, 1934. 52, inc. ads.   P

SEKON, G. A., *pseud.* [George Augustus Nokes].
Locomotion in Victorian London.
London: Oxford Univ. Press, 1938. Frontis., xi, 211, [27] pl.; illd.   B
     53–75 on Thames passenger boats.

SENIOR, William, 1839?–1920.
The Thames from Oxford to the Tower.
London: Nimmo, 1891. Frontis., xii, 120, [29] pl.   B

SHARP, James (ed.).
Extracts from Mr. Young's six months tour through the north of England, and from the letter of an un-
known author, published in the London Magazine, for October, 1772, on the subject of canal navigations.
[London?]: [?], 1774. iv, 29, [i], [7] pl., some fldg.   B

SHARPNESS DOCKS & GLOUCESTER & BIRMINGHAM NAVN. CO.
The port of Gloucester: 'The Ocean Port for the Midlands': shipping, manufacturing, mining.
Gloucester: SDGBN, 1936. 64; illd., ads.   T

SHAW, S[tebbing], 1762–1802.
A tour to the West of England in 1788.
London: Robson & Clarke, 1789. viii, 602.   B
     191–201 Wye trip; 252–3 Sapperton Tunnel; 538–48 Basingstoke Canal.

SHOWELL, Charles.
Shakespeare's Avon from source to Severn.
Birmingham: Cornish, 1901. [vi], 207, fldg. map; illd.   B
   2nd edn. 1925.

SOUTHEY, Robert, 1774-1843.
Journal of a tour in Scotland in 1819.
London: Murray, 1929. Frontis., li, 276, [1] pl.   B
      167-207 on Caledonian Canal.

STRETTON, Clement E[dwin].
The stone roads, canals, edge-rail-ways, Outram-ways, and electric rail-ways in the county of Leicester.
Leicester: Adam Bros. & Shardlow (printer), 1907. [i], 237-247.   B
   Extract from British Assn's. 'Guide to Leicester and district'.

SWINNERTON-BEET, A. (compiler).
Aire & Calder Navigation and port of Goole.
Leeds: ACN, 1937. 120; maps, illd.   P

TAYLOR, John, 1580-1653.
Taylor on Thame Isis: or the description of the two famous rivers of Thame and Isis . . . also a
discovery of the hinderances which doe impeach the passage of boats and barges . . .
London: Haviland (printer), 1632. [viii], 20.   B,N
   Imprint cropped off B copy.

THACKER, Fred[erick] S[amuel].
Kennet country.
Oxford: Blackwell, 1932. 387; drgs., map endpprs.   B

THACKER, Fred[erick] S[amuel].
The stripling Thames: a book on the river above Oxford.
London: Thacker, 1909. Frontis., viii, 495, [42] pl.; drgs., maps.   B

THACKER, Fred[erick] S[amuel].
The Thames highway: a history of the inland navigation.
London: Thacker, 1914. Map frontis., [viii]. [4] pl.   B
   Facs. edn. by David & Charles (Newton Abbot) 1968 as 'The Thames highway, Vol. I: General history'.

THACKER, Fred[erick] S[amuel] .
The Thames highway: a history of the locks and weirs.
London: Thacker, 1920. 525, [17] pl.; maps.   Ḃ
   Facs. edn. by David & Charles (Newton Abbot) 1968 as 'The Thames highway: Vol. II: Locks
   and weirs'.

The Thames and its story: from the Cotswolds to the Nore.
London: Cassell, 1906. Frontis., [viii], 376, 8 (pub. cat.), [47] pl.; maps.   B
   2nd edn. in 2 vols. 1910.

THAMES RIVER USERS' ASSN.
The great highway.
London: TRUA, [c. 1948] . [20]; illd., maps.   T

THOMAS, Richard.
Observations and directions for navigating the Severn, accompanied with several useful tables.
Falmouth: Brougham (printer), 1816. 16.   C

Tombleson's Thames. See FEARNSIDE, W. G.

TRACY, W[illiam] Burnett.
Port of Manchester: a sketch of the history and development of the Manchester Ship Canal . . .
Manchester: Hind et al., 1901. [vi], 110.   B

TRENT NAVIGATION CO.
The Trent: a highway of commerce between the ports on the Humber and the industrial centres of
the Midlands.
Nottingham: TN, [c. 1938]. 16; illd.   P

VANDERSTEGEN, William.
The present state of the Thames considered; and a comparative view of canal and river navigation.
London: Robinson, 1794. [iii], 76.   B

WAKEMAN, William Frederick, 1822-1900.
Lough Erne, Enniskillen, Belleek, Ballyshannon, and Dundoran, with routes from Dublin to
Enniskillen and Dundoran, by rail or by steamboat.
Dublin: Mullany, 1870. [i], vi, 165, [1] (ad.).   B

WALKER, James.
The Tyne as a navigable river.
[London?]: [Inst. Civil Engineers?], 1905. 14, [14], fldg. plan.   T
    Text of address to Newcastle Assn. of I.C.E. students.

WALLS, Ernest.
The Bristol Avon.
Bristol: Arrowsmith, 1927. Frontis., 318, [11] pl.; drgs., map endpprs. (Rivers of England Series). B

WALLS, Ernest.
The Salisbury Avon.
Bristol: Arrowsmith, 1929. Frontis., xii, 288, [10] pl.; drgs., map endpprs. (Rivers of England
Series).   B

WALLS, Ernest.
Shakespeare's Avon.
Bristol: Arrowsmith, 1935. Frontis., 320, [13] pl.; drgs., map endpprs. (Rivers of England Series). B

WALMSLEY, Clement.
History of the Manchester Ship Canal, with considerations as to its future prospects.
Manchester: Heywood, 1887. 65.   B

WARDALE, Harry.
The Weaver Navigation.
Manchester: Rawson, 1935. 11.   T
    Reprinted from *Trans. Lancs. Chesh. Antiq. Soc.*, **49**.

WATERS, Brian.
Severn stream.
London: Dent, 1949. Frontis., 206, [15] pl.   B

WATERS, Brian.
Severn tide.
London: Dent, 1947. Frontis, vii, 183, [15] pl.   B
    2nd edn. 1955.

WEAVER NAVN. TRUSTEES.
River Weaver Navigation . . .
Northwich, Cheshire: WNT, [1939]. 16; illd., maps.   P

WELLS, Samuel.
A collection of the laws . . . of the Bedford Level Corporation.
London: for the author, 1828. xxii, 802.   B
    Although issued before Vol. 1, this is Vol. 2 of 'The history of the drainage'.

WELLS, Samuel.
The history of the drainage of the great level of the Fenns called the Bedford Level; . . .
London: for the author, 1830. xviii, 832.   B
    Although issued after Vol. 2 (*see* preceding) this is Vol. 1.

WHEELER, W[illiam] H[enry], 1832-1915.
A history of the Fens of south Lincolnshire, being a description of the rivers Witham and Welland
and their estuary . . .
Boston: Newcomb, 1868. 188, fldg. map.   B
    2nd edn. (*c.* 1897) 'greatly enlarged'.

WHITE, Archie.
Tideways & byways in Essex & Suffolk.
London: Arnold, 1948. 216; illd., maps, map endpprs.   B
    2nd imp. 1948.

WILLAN, T[homas] S[tuart] (ed.).
The navigation of the Great Ouse between St Ives and Bedford in the seventeenth century.
Streatley, Beds.: Bedfordshire Historical Record Soc., 1946. [v], fldg. map, 153. (BHRS Vol. 24).   B

The Wye tour, comprising a few historical and geological notes on the river from Plinlimmon to Chepstow, . . .
Ross: Hill, [1860]. iv, 108; [8] ads., fldg. map; illd.   B
    New enlarged edn. c. 1880.
       5–7 on navigation.

YOUNG, Arthur, 1741–1820.
A six months tour through the North of England.
London: Strahan et al., 1770. 4 vols. [iii], iv, iii–xxv, 401, 7 pl., fldg. table, [3] (pub. ad.); [i], vi,
502, 6 pl.; vii, 440, 12 pl.; viii, 594, [1] pl.   B
    2nd edn., 4 vols., 1770–1.
       III: 251–291 and pl. 4–10 re Bridgwater Canal.
       See also SHARP, James.

YOUNG, T[homas] M.
Manchester and the Atlantic traffic.
London: Sherratt & Hughes, 1902. Fldg. plan frontis., ix, 88, [20] pl.   B

3: Plans and proposals

ALLNUTT, Zachariah.
Considerations on the best mode of improving, the present imperfect state of the navigation, of the River Thames from Richmond to Staines . . .
Henley: for the author, 1805. 46, [2] fldg. maps, [1] pl.; drg.   B

BADESLADE, Thomas.
The new canal, intended for improving the navigation of the city of Chester, . . . compared with the Welland, . . .
Chester: Adams, 1736. 22, [1] pl.   B

BADESLADE, Thomas.
Reasons humbly offer'd . . . shewing how the works now executing . . . to recover and preserve the navigation of the River Dee, will destroy the navigation; . . .
Chester: Adams, 1735. 14, [6] maps.   B

BADESLADE, Thomas.
A scheme for draining the great level of the Fens, called Bedford-Level; and for improving the navigation of Lyn-Regis . . .
London: Roberts, 1729. Map, 10.   C
    'No more than 150 copies printed'. Possibly issued with 2 maps.

BIRMINGHAM & BRISTOL CHANNEL IMPROVED NAVIGATION COMMITTEE.
Birmingham & Bristol Channel improved navigation.
Birmingham: BBCINC, 1887. 34, [4] fldg. pl.   B
    Proposes 200-ton waterway throughout.

BOTTERILL, W. J.
Proposed naval base between Norwich & Yarmouth, and ship canal across England (North Sea to Bristol Channel).
Norwich: East of England Newspaper Co., 1908. 8; maps, illd.   C
    Reprinted from The Norfolk Chronicle, 23 October 1908.

BOTTERILL, W. J.
Proposed Norwich to Yarmouth Ship Canal . . . [and] . . . ship canal across England . . .
Norwich: Norwich Mercury, [1909?]. [6]; illd.   C
    Reprinted from *Black and White*.

BRAND, Alexander.
An abstract and true state of the undertaking . . . for making a navigation from Leith to Edinburgh . . .
[?]: [?], [*c.* 1720]. [2].   B

BROWN, Crisp.
The speech of Alderman Crisp Brown . . . Tuesday, the eighth day of Sept. 1818 . . .
Norwich: Burks & Kinnebrook (printers), [1818]. 24, fldg. plan.   B
      Re making Yare navigable to Norwich.

BROWNE, John.
The benefits which arise to a trading people from navigable rivers. To which are added . . . a scheme,
for the establishment of a company, to make the River Shannon navigable . . .
Dublin: Ewing *et al.*, 1729. vi, 45.   B

BUNGE, J[ulius] H[enri] O[tto].
Dam the Thames: a plan for a tideless river in London.
London: Thames Barrage Assn., 1935. Fldg. leaf, 8; illd.   T

BUNGE, J[ulius] H[enri] O[tto].
Tideless Thames in future London.
London: Thames Barrage Assn., 1944. 122; illd., map endpprs.   B

CHAPMAN, William, 1749-1832.
Address to the subscribers to the canal from Carlisle to Fisher's Cross.
Newcastle: Walker, 1823. 16.   B

CHAPMAN, William, 1749-1832.
Facts and remarks relative to the Witham and the Welland . . . observations . . . on the means of
improving the channel of the Witham . . .
Boston: Hellaby (printer), 1800. viii, 75.   B

CHAPMAN, William, 1749-1832.
Observations on Mr. John Sutcliffe's report on a proposed line of canal, from Stella to Hexham . . .
Newcastle: Whitfield, 1797. 18.   N

CHAPMAN, William, 1749-1832.
Observations on the advantages of bringing the Grand Canal round by the circular road into the
River Liffey.
Dublin: Byrne, 1785. [i], 29.   N

CONDER, Francis R.
The actual and the possible cost of conveyance between Manchester and Liverpool.
Manchester: Heywood, 1882. 29-44.   C
    Paper delivered to Manchester Statistical Soc.
      Advocates ship canal to Manchester.

[CONDER, Francis R.]
Inland transport and the Manchester Ship Canal.
Paisley: Gardner, 1886. 32.   C
    Reprinted from the *Scottish Review*.

CONGREVE, Thomas.
A scheme . . . for making a navigable communication between the Rivers of Trent and Severn, in the
County of Stafford.
London: Curll, 1717. 15.   B
    Another edn. by Shuckburgh (London) 1753, with fldg. map.

CONVENERY OF THE TRADES OF LEITH.
Resolutions . . . respecting the proposed Union Canal.
Leith: Gilchrist & Heriot (printers), 1816. 14.   B

COOK, W. B., and WILLOUGHBY, F. (patentees).
Improved method of utilizing canals for traffic and means therefor; designed by W. B. Cooke, C.E., to restore prosperity to the canals in the United Kingdom.
[Daventry?]: [Cook & Willoughby?], 1896. 14.   C
    Advocates converting inter-urban canals into railways.

CRAWFORD, J[ames] Law.
Forth and Clyde Ship Canal: in relation to the development of commerce.
Glasgow: Maclaren, [1891]. vii, 5–108, III fldg. pl.; diag.   B

CUBITT, W.
Description of a plan for a Central Union Canal, which will lessen the distance and expense of canal navigation between London and Birmingham, and unite the Birmingham–Warwick and Birmingham–Coventry- and Oxford Canals . . .
London: Roake & Varty. 1832. 7, fldg. map.   C
  Another edn. 1833 with 2 fldg. maps.

CUNDY, N[icholas] W[ilcox].
Imperial Ship Canal from London to Portsmouth. Mr. Cundy's reply to anonymous and other authors of malignant abuse and misrepresentation, on his projected line, furnishing truth for libel . . .
London: Rivington, 1828. 22, fldg. map.   B

DAWSON, James.
Canal extensions in Ireland . . .
Dublin: Porter (printer), 1819. Fldg. map frontis., [vii], 48.   C

Deep water railway docks in the Mersey, versus Manchester Ship Canal. By Progress.
Edinburgh: Crawford & M'Cabe (printers), 1887. 12, [3] plans.   B

A designe for bringing a navigable river from Rickmansworth . . .
See FORDE, Edward.

DUMBELL, Mr. [John?].
Manchester Ship Canal. Mr. Dumbell's observations relative to making Manchester and Warrington into sea ports . . .
London: Richardson, 1826, 2nd edn. 23.   C
  1st edn. not seen.

[EDINBURGH & GLASGOW UNION CANAL SUBSCRIBERS.]
Correspondence between the committee of subscribers to the proposed Union Canal between Edinburgh and Glasgow, and the Right Hon. the Lord Provost and Magistrates of Edinburgh . . .
Edinburgh: Oliver & Boyd, 1814. 15.   C

[EDINBURGH & GLASGOW UNION CANAL SUBSCRIBERS.]
Observations [by the committee of subscribers of the EGUC].
Edinburgh; [?], 1814. 16.   C

[EDINBURGH & GLASGOW UNION CANAL SUBSCRIBERS.]
Observations by the Union Canal Committee, on the objections made by the inhabitants of Leith to this undertaking.
[Edinburgh]: Oliver & Boyd (printer), [1817]. 31, 7.   C

EDWARDS, J.
Essay on a proposed canal, to unite the rivers Itching and Wey, which will make it completely navigable from the Thames to Southampton Bay . . .
[?]: [?], [1795?]. 15.   G

EWART, John.
[A letter to Charles Dundas, Chairman of the Committee of Management of the Kennet and Avon Canal, setting out Dr. Ewart's arguments that the canal will not affect the hot springs of Bath.]
[Bath?]: [KAC?], 1796. 4.   C

[FORDE, Edward, 1605–1670.]
A designe for bringing a navigable river from Rickmansworth in Hartfordshire, to St Gyles in the Fields. . .
London: for Clarke, 1641. 11; map.   B
  Another edn. 1720.

FOX, J. J.
A letter to 'An Inhabitant of Stamford', exposing the futility of his remarks on A Letter addressed
to J. J. Oddy, Esq. and on the projected Harborough and Stamford Junction Canal . . .
London: Turner & Harwood (printer), 1811. 39, errata slip.   C

FOX, J. J.
A letter to Joshua Jepson Oddy, Esq. shewing that the advantages . . . from a canal from Stamford
to Harboro', will be . . . greater . . . than . . . from the Stamford Junction Canal, proposed by Mr.
Oddy.
Stamford: Newcomb (printer), 1810. 49.   B

GRAY, J[ohn], 1724–1811.
Reflections on inland navigations: and a new method proposed for executing an intended navigation
betwixt the Forth and the Clyde . . .
London: Cadell, 1768. 48.   B

GRUNDY, John (senior), and GRUNDY, John (junior).
A scheme for the restoring and making perfect the navigation of the River Witham from Boston
to Lincoln . . .
[?]: [?], 1744. [i], 48.   B

GRUNDY, John (junior), c. 1719–1783.
A scheme for executing a navigation from Tetney-Haven to Louth . . . to which is added the report
of John Smeaton . . . [on] . . . a scheme of navigation from Tetney Haven to Louth . . .
Nottingham: Parker, 1761. 24.   N

HADFIELD, E[llis] C[harles] R[aymond], b. 1909.
Canals between the English and the Bristol Channels.
[?]; [?], 1942. 59–47.   P
    Reprinted from *The Economic History Review*.

[HARVEY, James W.]
The proposed Manchester Ship Canal: facts and figures in favour of a tidal navigation to Manchester
. . . by Mancuniensis.
Manchester: Heywood, 1882. 35.   B
    Revd. edn. 1882, with fldg. map, under slightly altered title.

HELY, James.
A modest representation of the benefits and advantages of making the River Avon navigable from
Christ-Church to the city of New Sarum . . .
London: for the author, 1672. 24.   B,N
    Title page of B copy cropped.

HIBBARD, John.
Hibbard's essay of the great utility and interest that might arise from a circular & other collateral,
&c. canal navigation and drainage . . .
London: for the author, [c. 1804] . 44.   B

HIBBARD, John.
Statements on the great utility of a circular and other internal &c. canal navigation, and drainage . . .
London: for the author, [c. 1800]. 63;   B

HIBBARD, John.
Thoughts . . . on the great utility of a circular and other inland canal navigation, in Britain . . .
London: for the author, 1798. 30.   B

HIRD, George S[mith].
The Sheffield & South Yorkshire Navigation. The Sheffield Canal. Paper read before the Sheffield
Society of Engineers, Sheffield, 19th April, 1890.
[Sheffield?]; [the author?], [1890?]. 29; plan.   B

Inland transport and the Manchester Ship Canal.
See CONDER, Francis R.

LLOYD, Samuel, b. 1827 [?].
Canal improvements between the four rivers. A national necessity.
London: Simpkin *et al.*, 1889. 2nd edn. Map, [2] pl., xii, 5–68.   C
    For 1st edn. *see* 'A national canal . . . '.

LLOYD, Samuel, b.1827 [?].
England needs inland steam navigation: a tale.
London: Pickering. 1885. xiii, 47–89.   B
    Cover title 'The General Election: a tale out of Egypt and what England needs'.
      Apparently Pt. III of a longer work but the only part to deal with inland navigation. Proposes
      large-gauge canals to link Birmingham to Mersey, Severn and Trent.

LLOYD, Samuel, b. 1829 [?].
A national canal between the four rivers a national necessity.
London: Hogg, 1888. Map frontis., 64.   B
    For 2nd edn. *see* 'Canal improvements . . . '.

LLOYD, Samuel, b. 1827 [?].
The proposed national canal.
London: Simpkin *et al.*, 1887. 32.   B
    To join the Thames to the Mersey and the Humber to the Severn.

London and Cambridge Junction Canal.
Minutes of the evidence taken before a committee of the House of Commons . . .
London: Philanthropic Soc. (printers), 1812. Fldg. map frontis., 451.   C

MACGREGOR, John.
Observations on the River Tyne, with a view to the improvement of its navigation; . . .
Newcastle-upon-Tyne: Charnley, 1832. vii, 118, [1] pl.   B

MACKELL, Robert, and WATT, James.
An account of the navigable canal, proposed to be cut from the River Clyde to the River Carron . . .
London: [?], 1767. Fldg. map., 18.   B

MATHEW, Francis.
A mediterranean passage by water, from London to Bristol, &c. and from Lynne to Yarmouth, and
so consequently to the City of York: for the great advancement of trade & traffique.
London: Newcomb (printer), 1670. [v]. 12.   B

MATHEW, Francis.
Of the opening of rivers for navigation, the benefit exemplified by the two Avons of Salisbury and
Bristol. With a mediterranean passage by water for billanders of thirty tun, between Bristol and
London.
London: Cottrel, 1655. [iv], 11, plan.   B
    2nd edn. 1656.

MATHEW, Francis.
To his Highness, Oliver, Lord Protector of the Common-wealth . . . is humbly presented: A mediter-
ranean passage by water between the two sea towns Lynn & Yarmouth, upon the two rivers the Little
Owse, and Waveney. With farther results. Producing the passage from Yarmouth to York.
London: Dawson (printer), 1656. 15.   B

MULLINS, Bernard.
Thoughts on inland navigation, with a map, and observations upon propositions for lowering the
waters of the Shannon, and of Lough Neagh, addressed to the Right Hon. E. G. Stanley, Chief
Secretary for Ireland.
Dublin: Hope (printer), 1832. 54, fldg. map.   P,N

NETTLEFORD, J[ohn] S[utton].
Garden cities and canals.
London: St Catherine Press, 1914. xiii, 220; map.   B
    Develops Royal Commission proposals for Cross.

ODDY, J[oshua] Jepson, d. 1814.
A sketch of the improvement of the political, commercial, and local interests of Britain, as exemplified by the inland navigations of Europe in general, and of England in particular; including details relative to the intended Stamford Junction Navigation . . .
London: Stockdale, 1810. vii, 143, 8 (pub. cat.), fldg. map.   B

PAGE, Frederick.
Observations on the present state, and possible improvement of the navigation and government of the River Thames.
Reading: Smart & Cowslade, 1794. Fldg. map, frontis., 41.   B

PEMBER, [Edward Henry], 1833–1911.
Manchester Ship Canal Bill. Session 1884. Reply of Mr. Pember, Q.C., on behalf of the promoters of the Bill, before the Select Committee of the House of Lords, 20th, 21st, and 22nd May, 1884.
London: for the promoters, [1884]. iv, 139.   B

PEMBER, [Edward Henry], 1833–1911.
Manchester Ship Canal. Speeches by Mr Pember, Q.C. (Counsel for the promoters), before the Select Committee of the House of Commons, 1st May; 3rd and 4th July, 1883 . . .
London: for the promoters, [1833]. 119.   B

PHILLIPS, John.
A treatise on inland navigation: illustrated with a whole-sheet plan, delineating the course of an intended navigable canal from London to Norwich and Lynn . . .
London: Hooper, 1785. xii, [6] (subscribers' list), [2] (pub. cat.), 50, fldg. map, pl.   B

POWNALL, J[ohn] F[rederick], 1900–1971.
The projected Grand Contour Canal: to connect with estuaries and canals in England.
Birmingham: Cotterell, 1942. 35; maps.   C

POWNALL, J[ohn] F[rederick], 1900–1971.
Transport reform in Great Britain.
London: the author, [1932]. [4], fldg. map.   B
       First publication of Grand Contour Canal proposals.

The proposed Manchester Ship Canal . . . By Mancuniensis.
See HARVEY, James W.

The proposed ship canal between Rotherham and Goole. Facts and figures in favour of water carriage for the great import and export trade of South Yorkshire and North Derbyshire. By F.R.
Rotherham: Garnett & Whitehead (printers), 1883. 35, fldg. map.   B

PROVAND, A[ndrew].
The Manchester Ship Canal scheme: a criticism.
Manchester: Heywood, 1883. 48.   B
     2nd (enlarged) edn. 1883, with plate and fldg map.

PROVIS, William Alexander, 1792-1870.
Suggestions for improving the canal communication between Birmingham, Wolverhampton, Shropshire, Cheshire, North Wales, and Manchester by means of a new canal from Middlewich to Altringham.
London: Weale, 1837. 12, fldg. plan.   N

PYKE, T.
Grand ship canal, from Bridgwater to Seaton . . . forming a junction of the English, with the Bristol, and Irish Channels.
Bridgwater: A'Court (printer), 1823. 16.   C

RICHARDSON, W[illiam], 1740-1820.
Memoir on the subjects of making the lower Bann navigable and restraining the winter redundance of Lough Neagh.
Belfast: Mackay (printer), 1811. 54.   G

RIDGE-BEEDLE, P[eter] D[enniston].
Report of Mid-Scotland Ship Canal Committee . . . The case for the canal . . .
Glasgow: Aird & Coghill (printers), 1944. 8.   B

RIVER AVON IMPROVEMENT ASSOCIATION.
The River Avon. Why its navigation should be restored and how it may be done.
Evesham: RAIA, 1905. [iv], 49, [2] maps; illd.   P

SHARP, James.
An address to the . . . Lord-Mayor . . . of the City of London, on the importance and great utility
of canals in general; and the advantages . . . expected from the canal now proposed to be made from
Waltham-Abbey . . .
[London?]: [for the author?], 1774. [i], 18.   B

SMEATON, John, 1724–1792.
A review of several matters relative to the Forth and Clyde Navigation, . . . with some observations
on the reports of Mess. Brindley, Yeoman, and Golburne.
[Edinburgh?]: FCN Co., 1768. [i], 34.   B

STEELE, Thomas, 1788–1848.
Practical suggestions on the general improvement of the navigation of the Shannon, between
Limerick and the Atlantic . . .
London: Sherwood et al., 1828. viii, fldg. chart, 151.   B

STEUART, Henry, 1759–1836.
Account of a plan for the better supplying of the city of Edinburgh with coal; together with an
examination of the merits of the two principal lines, pointed out for the intended canal between
Edinburgh and Glasgow.
Edinburgh: Bell & Bradfute, 1800. 142.   P,N
        This started an exchange with James Dunlop (see two following entries and 'Observations'
        and 'Remarks' in Appendix).

STEUART, Henry, 1759–1836.
Letter, addressed to James Dunlop, esq. on his late remarks on the proposed plan for the better
supplying the city of Edinburgh with coal . . .
Edinburgh: Hill, 1801. 12.   N

STEUART, Henry, 1759–1836.
Supplement to an account of a plan, for the better supplying the city of Edinburgh with coal; comprising an examination of an anonymous pamphlet lately published, under the signature of An old
coalmaster.
Edinburgh: Longman & Rees, 1800. [iv], 204.   B

STOKES, H[enry] P[aine], 1849–1931.
Better canals, better trade. Remarks on the carriage of heavy goods, raw material, and agricultural
products.
Wolverhampton: Barford & Newitt (printers), 1885. 23.   P
        Proposes enlargement of important canals.

STRATFORD, Ferdinando.
A plan for extending the navigation from Bath to Chippenham. In a letter to the worshipful society
of merchants in the city of Bristol.
Bristol: Farley (printer), 1765. 18.   N

TATHAM, William, 1752–1819.
Plan for improving the city of London by means of navigable canals & commercial basons.
[London?]; [?], 1799. 163.   N

TAYLOR, John, 1580–1653.
A new discovery by sea, with a wherry from London to Salisbury . . .
London: for the author, 1623.   N
    Reprinted in HINDLEY, Charles (ed.). The Old Book Collector's Miscellany, Vol. 3, 1873.
    Original not seen.
        Advocates improvement of navigation of Avon.

THAMES BARRAGE ASSOCIATION.
Dam the Thames: a matter of sanitation, comfort and economy. A plan for a tideless river in London.
Effingham, Surrey: TBA, 1936 (3rd edn.). 16; illd., maps.   P
   Only edn. seen.

THAMES BARRAGE COMMITTEE.
The port of London and the Thames barrage.
London: Swan Sonnenschein, 1907. v, 193, [23] pl. (10 fldg.); diags.   B

THOMAS, Richard.
Hints for the improvement of the navigation of the Severn, comprising information which may be applicable to other navigations.
Falmouth: Brougham (printer), 1816. 36.   C

THOMAS, William.
Observations on canals and rail-ways, illustrative of the ... advantages to be derived from an iron rail-way ... between Newcastle, Hexham, and Carlisle ... also, second edition, report of Barrodall Robert Dodd ... on a proposed navigable canal between Newcastle and Hexham ...
Newcastle-upon-Tyne: Heaton *et al.*, 1825. 52.   B

THOMPSON, Jonathan.
Observations on the most advantageous line of country through which a canal navigation may be carried, from Newcastle-upon-Tyne, or North Shields, towards Cumberland &c. ...
Newcastle: Sands, 1795. 24.   B

THOMPSON, William.
A letter to the subscribers to the intended Stamford Junction Navigation ...
Stamford: Drakard, 1810. 37.   B

UPTON, John.
Observations on the Gloucester & Berkeley Canal ...
Gloucester: Walker (printer), 1815. 43.   C

WARD, John.
Observations on the advantages of the proposed Western Canal, from Newbury to Bath.
Marlborough: [for the author?], 1790. 3.   G

WATT, James, 1736–1819.
A scheme for making a navigable canal from the city of Glasgow to the Monkland Coalierys.
[Glasgow?]: [?], [*c.* 1770?]. 12.   B

WHITWORTH, Richard.
The advantages of inland navigation: or some observations offered to the Public, to shew that an inland navigation may be easily effected between the three great ports of Bristol, Liverpool, and Hull; together with a plan for executing the same.
London: Baldwin, 1766. Fldg. frontis., vi, 74, [4] fldg. tables, fldg. plan.   B
   Corrected edn. 1766.

WHYMAN, Thomas.
An address, to the Leeds & Liverpool Canal Company, on the proposed deviation of the line of that canal, by Blackburn, Chorley, &c.
Preston: Walker (printer), 1792. 44.   B

WILLIAMS, Charles Wye, 1779–1866.
A speech on the improvement of the Shannon, being in continuation of the debate in the House of Commons ...
London: Bain, 1835. 60, fldg. maps.   N

YARRANTON, Andrew, 1616–1648 [?].
England's improvement by sea and land . . . making the great rivers of England navigable . . .
London: for the author, 1677–81. 2 pts. Licence leaf, [xviii], 195, [8] fldg. pl.; [viii], 212, [7] fldg. pl.
>   Mispaginations: Pt. 1: 110 for 112, 73–96 omitted; Pt. 2: 121–8 omitted. N also records other mispaginations.
>   Another edn. by Parkhurst (London) 1698.

LICENSED.

Octob. 4.
1676.

Roger L'estrange.

ENGLAND'S
𝔍𝔪𝔭𝔯𝔬𝔟𝔢𝔪𝔢𝔫𝔱
BY
SEA and LAND.
TO
Out-do the *Dutch* without Fighting,
TO
Pay Debts without Moneys,
To set at Work all the POOR of *England* with the
Growth of our own Lands.
To prevent unnecessary SUITS in Law;
With the Benefit of a Voluntary REGISTER.
Directions where vast quantities of Timber are to be had
for the Building of SHIPS;
With the Advantage of making the Great RIVERS
of *England* Navigable.
RULES to prevent FIRES in *London*, and other Great CITIES;
With Directions how the several Companies of Handicraftsmen in *London*
may always have cheap Bread and Drink.

By *ANDREW YARRANTON*, Gent.

LONDON,
Printed by R. *Everingham* for the Author, and are to be sold by *T. Parkhurst*
at the Bible and three Crowns in *Cheap-side*, and *N. Simmons* at the Princes
Arms in S. *Paul's* Church-yard, M DC LXXVII.

Figure 7.1. Seventeenth-century optimism from Andrew Yarranton.

4: *Waterway engineering; engineering biography*

AFFLECK, Thomas.
Patent hydraulic deepener . . . [for] deepening and clearing out of rivers, and . . . the cutting and preserving of new channels in friths, bays, and arms of the sea.
Glasgow: [for author?], 1833. 27, 9 pl.   B

ALLEN, [Charles] Grant [Blairfindie].
Biographies of working men.
London: Soc. Promoting Christian Knowledge, 1884. 191, 4 (pub. cat.).   P,(B)
>   5–29 on 'Thomas Telford, stonemason'.

BEARDMORE, Nathaniel, 1816–1872.
Hydraulic tables, to aid the calculation of water and mill power . . . drainage & navigable rivers . . .
London: Waterlow, 1850. [i], xxxvi, 60.   C
   2nd edn. 1852.
      Much of this material later incorporated into 'Manual of hydrology' (q.v.).

BEARDMORE, Nathaniel, 1816–1872.
Manual of hydrology: containing I.—Hydraulic and other tables. II.—Rivers, flow of water, springs, wells, and percolation. III.—Tides, estuaries, and tidal rivers. IV.—Rainfall and evaporation.
London: Waterlow, 1862. xii, 384, [20] pl. (16 fldg.).   B
   Further edns.: 1872 Waterlow (London); 1906, 1914 Spon (London).
      Includes lock gate design.

BELLASIS, E[dward] S[kelton], b. 1855.
River and canal engineering.
London: Spon, 1913. x, 215, 48 (pub. cat.); illd.   B
   3rd edn. 1931.

BROOKS, William Alexander.
Treatise on the improvement of the navigation of rivers: with a new theory on the cause of the existence of bars.
London: Weale, 1841. v, 154.   B

BURNESS, William.
Practical river reform: in drainage and navigation, in water power and irrigation, in warping land and storing water.
London: Bradbury *et al.*, 1882. 31.   B

CALVER, Edward Killwick.
The conservation and improvement of tidal rivers, considered principally with reference to their tidal and fluvial powers.
London: Weale, 1853. x, 101, fldg. plan; diags.   B

CHAPMAN, William, 1749–1832.
Observations on the various systems of canal navigation . . . in which Mr. Fulton's plan of wheel-boats and the utility of subterraneous and of small canals are particularly investigated, including an account of the canals and inclined planes of China.
London: Taylor, 1797. [vii], 104, [1] (pub. cat.), 3 pl., fldg. map.   B

CHAPPELL, Metius.
British engineers.
London: Collins, 1942. 48, [8] pl.; illd. (Britain in Pictures Series).   B

CRESSY, Edward.
Stories of engineering adventure: railways — ships — canals.
London: Warne, 1928. Frontis., x, 214, [23] pl.; maps.   B
     154–160, 167–177 on British canals.

DARBY, H[enry] C[lifford].
The draining of the Fens.
Cambridge: University Press, 1940. Frontis., xix, 312, [26] pl.; maps, diags.   B
   2nd edn. 1956; 2nd imp. 1968.

DEAS, James.
The River Clyde: an historical description of . . . the harbour of Glasgow, and of the improvement of the river from Glasgow to Port-Glasgow.
Glasgow: MacLehose, 1876. xi, 127, 5 fldg. pl.; fldg. tables.   B

DICKINSON, H[enry] W[inram], 1870–1952.
Robert Fulton, engineer and artist: his life and works.
London: Lane, 1913. Frontis., xv, 333, 16 (pub. cat.), [21] pl.   B

ERSKINE, Robert.
A dissertation on rivers and tides . . . in particular . . . the River Thames.
London: for the author, 1770. 24.   B
   2nd edn. 1780.

FLOWER, Lamorock.
The River Lee 'up to date'.
London: Mitton, 1893. Fldg. map frontis., 39.   B
   Revd. version of paper of 1887.
     Largely relates to public health aspects.

FORBES, U[rquhart] A[twell] and ASHFORD, W. H. R.
Our waterways: a history of inland navigation considered as a branch of water conservancy.
London: Murray, 1906. xv, fldg. map, 336.   B

FRISI, Paolo, 1728–1784.
A treatise on rivers and torrents; with the method of regulating their course and channels . . . to
which is added an essay on navigable canals . . . translated by Major-General John Garstin . . .
London: Longman et al., 1818. vi, xx, 184, [2] fldg. pl., fldg. table.   B
  Originally pub. in Italy in 1762. New edns. by Weale (London) 1861, 2 vol., and Virtue (London)
  1868 and c. 1872.

FRITH, Henry.
The romance of engineering; stories of the highway, the waterway, the railway, and the subway.
London: Ward et al., [1892]. Frontis., xii, 364, [7] pl.; illd.   B
   2nd edn. 1895.
     87–154 largely on British waterways.

FULTON, R[obert], 1765–1815.
A treatise on the improvement of canal navigation; exhibiting the numerous advantages to be derived
from small canals. And boats . . . containing from two to five tons . . . with thoughts on, and designs
for, aqueducts and bridges of iron and wood.
London: Taylor, 1796. xvi, 144, [1] (pub. cat.), 17 pl.   B
  2 issues, both 4to: ordinary (page size c. 280 x 220 mm.), large (c. 293 x 233).

GIBB, Alexander.
The story of Telford: the rise of civil engineering.
London: Maclehose, 1935. Frontis., xxi, 357, [1] (pub. cat.), [25] pl., [3] fldg. maps.   B

GOWER, Francis Leveson (afterwards Francis Egerton, Earl of Ellesmere), 1800–1857.
Essays on history, biography, geography, engineering, &c., contributed to the 'Quarterly Review',
by the late Earl of Ellesmere.
London: Murray, 1858. vi, 473.   B
     201–51 on aqueducts and canals.

GRAHAME, Thomas.
Description of a plan for passing the clivities on canals, with little or almost no expenditure of water.
Glasgow: Lang (printer). 1836. [i], 14, 4 pl.   C
     Grahame's patent lift.

GRAHAME, Thomas.
A letter addressed to Nicholas Wood, Esq. on that portion of chapter IX of his treatise on railroads,
entitled, 'Comparative performances of motive power on canals and railroads'.
Glasgow: Smith, 1831. 40.   B

HUNTER, W[illiam] Henry, 1849–1917.
Dock and lock machinery: a technical manual.
London: Constable, 1921. xv, 207, 10 fldg. pl.; illd.   B

HUNTER, W[illiam] Henry, 1849–1917.
Rivers and estuaries, or streams and tides.
London: Longmans et al., 1913. [vii], 69; diags.   B

Industry illustrated; a memoir of Thomas Jackson, of Eltham Park, Kent . . .
London: for private circulation, 1884. Frontis., vii, 109, 40 pl. (some fldg.).   P
    Engineering contractor on several canals, including Caledonian.

JOHNSTONE-TAYLOR, F[rank].
River engineering: principles and practice.
London: Lockwood, 1928. xvi, 119, [1] (pub. cat.); illd.   B
    2nd edn. by Technical Press (London) 1938.

LESLIE, James.
Description of an inclined plane . . . on the Monkland Canal, at Blackhill, near Glasgow.
Edinburgh: Neill (printer), 1852. 16, pl. VII–IX.   B
    Reprinted from *Trans. Royal Scot. Soc. Arts.*

MACNEILL, John [Benjamin], 1793[?]–1880.
Canal navigation: on the resistance of water to the passage of boats upon canals . . . being the results
of experiments. (*See* illustration following page).
London: Roake & Varty, 1833. [v], 55, 7 pl. (some fldg.).   P,N

MANCHESTER SHIP CANAL CO.
Barton Aqueduct.
Manchester: MSC, [*c.* 1930]. 11; illd.   T

MINIKIN, R[obert] C. Royal.
Practical river and canal engineering.
London: Griffin, 1920. Frontis., vii, 123, XII pl., fldg. map; diags.   B

PATERSON, Malcolm McCulloch.
Pollution of the [Rivers] Aire and Calder: how to deal with it.
London: Spon, 1893. 32.   B
    Reprinted from *Engineering.*

PICKARD, A.
Pickard's system of canal transport.
[London]: Pickard, 1889. [8]; illd.   P

RENNIE, John, 1794–1874.
Autobiography of Sir John Rennie, F.R.S. . . .
London: Spon, 1875. viii, portrait, 464.   B

SANER, J[ohn] A[rthur].
Waterways, being the Vernon-Harcourt Lectures . . . 1913 . . . and other papers . . .
[?]: [?], [*c.* 1914]. 95.   B

SMILES, Samuel, 1812–1904.
James Brindley and the early engineers.
London: Murray, 1864. xv, 320, 8 (reviews), 32 (pub. cat.); illd.   B
    Abridged from 'Lives of the engineers'.

SMILES, Samuel, 1812–1904.
The life of Thomas Telford . . .
London: Murray, 1867. Frontis., xiii, 331, 6 (pub. cat.), [9] pl.; illd.   B
    A new edn. of the 'Life' originally pub. in 'Lives of the engineers'.

SMILES, Samuel, 1812–1904.
Lives of the engineers, with an account of their principal works; comprising also a history of inland
communication in Britain.
London: Murray, 1861–2. 3 vol. Frontis., xvii, 484, [1] pl., 12 (pub. cat.); illd. Frontis., xiv, 502,
[2] pl.; illd. Frontis., xxiii, 512, [1] pl.; illd.   B
    New edn. 1874 (5 vol.).   Popular edn. 1904 (5 vol.).
    Facs. edn. by David & Charles (Newton Abbot) 1968.

SMITH, Edward.
A description of the patent perpendicular lift, erected on the Worcester and Birmingham Canal,
at Tardebig, near Bromsgrove.
Birmingham: Smith, 1810. Frontis., 16, [2] pl.   C

CANAL NAVIGATION.

ON

Mitford

THE RESISTANCE OF WATER

TO THE PASSAGE OF

BOATS UPON CANALS,

AND OTHER

BODIES OF WATER,

BEING

THE RESULTS OF EXPERIMENTS,

MADE BY

JOHN MACNEILL, M.R.I.A.

MEMBER OF THE INSTITUTION OF CIVIL ENGINEERS, LONDON.

—————————*"mare per medium, fluctu suspensa tumenti*
Fert iter."   VIRGIL.

LONDON:

ROAKE AND VARTY, 31, STRAND.

GEORGE, BATH ; GORE, LIVERPOOL ; WRIGHTSON, BIRMINGHAM ; HODGES & CO.
DAME STREET, DUBLIN ; SMITH, GLASGOW.

1833.

Figure 7.2. Nicholas Wood's own copy of MacNeill's report on experiments performed on a 'quick boat' purchased from the Paisley Canal Company. This was renamed the 'Grahame and Houston' and tested on the Grand Juntion Canal. (*See* entry MACNEILL, previous page).

STEVENSON, David, 1815-1886.
Canal and river engineering, being the article 'Inland Navigation', from the eighth edition of the Encyclopaedia Britannica.
Edinburgh: Black, 1858. Fldg. map frontis., viii, 165, [2] (pub. ad.); drgs.   B
   2nd (1872) and 3rd (1886) edns. entitled 'The principles and practice of canal and river engineering . . .'.

STEVENSON, David, 1815-1886.
Remarks on the improvement of tidal rivers.
London: Weale, 1845. 36, [1] fldg. pl.; illd.   N
   2nd (enlarged) edn. 1849.

SUTCLIFFE, John.
[A letter] To the proprietors of the Somersetshire Coal Canal [re the means of transferring coal from the upper to the lower level].
Bath: [SSC?], 1800. 4.   C

TELFORD, Thomas, 1757-1834.
Life of Thomas Telford, civil engineer, written by himself . . . with a folio atlas of copper plates. Edited by John Rickman . . .
London: Payne & Foss, 1838. xxiv, 719, [1] pl.; illd. Atlas: frontis, [iv], 83 pl. (some fldg.).   B
   Atlas issued without plate 28.

Telford and Brindley: the story of their lives and engineering triumphs in roads, bridges and canals.
London: Chambers, [c. 1895]. 128; illd.   B

THOMAS, Gordon [Cale].
The 'Thomas' lift constructed at Foxton, Leicester, by the Grand Junction Canal Company.
London: Cook & Hammond, [c. 1906]. [20]; illd., map.   ?
   Facs. edn. by Danetre Press (Daventry) [c. 1978]. Orig. not seen.

VALLANCEY, Charles, 1721-1812.
A treatise on inland navigation, or, the art of making rivers navigable, of making canals in all sorts of soils, and of constructing locks and sluices.
Dublin: Ewing, 1763. [ii], x, 179, XXIV fldg. pl.   B

VERNON-HARCOURT, L[eveson] F[rancis], 1839-1907.
Achievements in engineering during the last half century.
London: Seeley, 1891. Frontis., viii, 311, [1] (pub. cat.); illd.   B
   2nd edn. 1892.
      206-269 on waterways.

VERNON-HARCOURT, Leveson Francis, 1839-1907.
Civil engineering as applied in construction.
London: Longmans et al., 1902, xvi, 624, 32 (pub. cat.); illd. (Longmans' Civil Engineering series.) B
   2nd edn. 1910; 2nd imp. 1919.
      323-436 on river and canal engineering.

VERNON-HARCOURT, Leveson Francis, 1839-1907.
Rivers and canals: the flow, control, and improvement of rivers, and the design, construction, and development of canals both for navigation and irrigation.
Oxford: Clarendon Press, 1896. 2 vols. xx, 342, X fldg. pl.; diags. xii, 343-704, XI-XIII fldg. pl.; diags.   B
      Extensively revised version of 'A treatise on rivers and canals'.

VERNON-HARCOURT, Leveson Francis, 1839-1907.
A treatise on rivers and canals relating to the control and improvement of rivers and the design, construction, and development of canals.
Oxford: Clarendon Press, 1882. 2 vols. xii, 352; diags. viii, 21 fldg. pl., 36 (pub. cat.). (Clarendon Press series).   B
      Extensively revised to become 'Rivers and canals'.

WEAVER NAVN. TRUSTEES.
Anderton boat lift, 1908.
[Northwich, Cheshire: WNT, c. 1908] . 13, IX pl.   ?
   Reprinted from *The Engineer* of 24 July 1908.
   Facs. edn. by Danetre Press (Daventry), n.d.
   Orig. not seen.

WHEELER, W[illiam] H[enry], 1832-1915.
Canals and inland navigation.
Boston: Newcomb, 1889. 32.   B
   Reprinted from *The Engineer*.

WHEELER, W[illiam] H[enry], 1832-1915.
Tidal rivers: their (1) Hydraulics, (2) Improvement, (3) Navigation.
London: Longmans *et al.*, 1893. viii, 467, 16 (pub. cat.); [2] fldg. pl.; illd. (Longmans' Civil
Engineering Series).   B

WHEELER, W[illiam] H[enry], 1832-1915.
The transporting power of water, as applied to the deepening & improvement of rivers, with a descrip-
tion of 'The Eroder'.
Boston: Newcomb, [c. 1892]. [4], 20, fldg. pl.; illd.   B
   'The Eroder' was a dredger designed by Wheeler.

WILLIAMS, Archibald, 1871-1934.
Victories of the engineer.
London: Nelson, [c. 1908]. 487; frontis., 259 illns.   B
   N records 11th edn. n.d.
      304-349 on waterways.

WOOD, Nicholas, 1795[?]-1855.
A practical treatise on rail-roads, and interior communication in general; . . . and tables of the comparative value of canals and rail-roads.
London: Knight & Lacey, 1825. [iii], 314, VI fldg. pl.   B
   2nd edn. by Hurst *et al.* (London) 1831; 2nd issue by Longman *et al.* (London) 1832. 3rd edn.
   by Longman *et al.* (London) 1838.

5: *Boats and boatmen; boatbuilding; boat propulsion and haulage*

BENHAM, Hervey [William Gurney], b. 1910.
Last stronghold of sail. The story of the Essex sailing-smacks, coasters and barges.
London: Harrap, 1948. 202; illd., map endpprs.   B

BENNETT, A[rthur] S[crivener].
Tide time.
London: Allen & Unwin, 1949. Frontis., [viii], 165, [32] pl.   B
      Thames sailing barges in trade and cruising.

BYFORD-JONES, W[ilfred].
Midland leaves: a travel notebook.
Wolverhampton: Midland News Assn., 1934. 214; illd.   P
      53-61 on boatmen at Wolverhampton.

CARR, Frank G[eorge] G[riffith], b. 1903.
Sailing barges.
London: Hodder & Stoughton, 1931. Frontis., 328, [48] pl. (1 fldg.); drgs.   B
   Revd. edn. 1951.

Conveyance upon canals & rail-roads compared.
London: Roake & Varty, 1831. 16.   C
   Another issue of 1831 has 17 pp., and 'By Detector' on title.
      On cost of traction.

FAIRBAIRN, W[illiam], 1789–1874.
Remarks on canal investigation, illustrative of the advantages of the use of steam, as a moving power on canals . . .
London: Longman *et al.*, 1831. 93, V fldg. pl.   B

GRAHAME, Thomas.
A letter, addressed to the proprietors and managers of canals and navigable rivers, on a new method for tracking and drawing vessels by a locomotive engine boat . . .
London: Longman *et al.*, 1825. [iv], 40.   C
   N records frontis.

GRAHAME, Thomas.
A letter to the traders and carriers on the navigations connecting Liverpool and Manchester . . .
Glasgow: Smith, 1833. viii, 3–29.   C
   2nd edn. 1834.
      Advocates the use of fast, iron boats.

GREEN, G[eorge] Colman.
The Norfolk wherry: its construction, evolution and history.
Wymondham, Norfolk: Reeve, 1953 (revd. edn.).
Frontis., xi, 50, [v], 12 (illns.).   B
   1st edn. by Model Yachting Assn., 1937, not seen.

HERBERT, A[lan] P[atrick], 1890-1971.
No boats on the river.
London: Methuen, 1932. Frontis., xi, 83, [23] pl.; maps, diags.   B
      Calls for new Thames waterbus service in London.

HODDER, Edwin, 1837–1904.
George Smith (of Coalville); the story of an enthusiast.
London: Nisbet, 1896. Frontis, 272, [8] (pub. cat.).   B
      The canal social reformer.

LANE, Leonard G.
Down the river to the sea: an historical record of the Thames pleasure steamers 1816 to 1934.
London: British Periodicals, [1934]. 112 (inc. ads.); illd.   B

LIVIUS, Barham, d. 1865.
A letter addressed to canal proprietors on the practicability of employing steam power on canals.
London: Hatchard *et al.*, 1842. Fldg. frontis., 24.   B

LONDON COUNTY COUNCIL.
A guide to the Council's river steamboat service.
London: LCC, 1905, 2nd edn. 78, fldg. map.   L
   1st edn. not seen.

McQUEEN, Andrew.
Clyde river-steamers of the last fifty years.
Glasgow: Gowans & Gray, 1923. Frontis., xii, 135, 47 pl.   B

McQUEEN, Andrew.
Echoes of old Clyde paddle-wheels.
Glasgow: Gowans & Gray, 1924. Fldg. frontis., xii, [31] pl.   B

MARCH, Edgar J.
Spritsail barges of Thames and Medway.
London: Marshall, 1948. Frontis., [viii], 304; illd.   B
   New edn. by David & Charles (Newton Abbot) 1970.

MARTIN, E[dward] G[eorge].
Sailorman.
London: Oxford Univ. Press, 1933. Frontis., [xi], 90, [5] pl.; drgs.   B
    Thames sailing barges.

MARX, Enid, & LAMBERT, Margaret.
English popular and traditional art.
London: Collins, 1946. 48, [8] pl.; illd. (Britain in Pictures series).   B
    38–40 and col. plate refer to narrow-boats.

O'CONNOR, John [Scorror].
Canals, barges and people.
London: Art & Technics, 1950. 96; illd., map endpprs.   B

PASK, Arthur T[homas].
The eyes of the Thames.
London: Ward & Downey, 1889. iv, 268, 16 (pub. cat.)   B
    32–43 describe journey on sailing barge on Regent's Canal and Thames.

POLLOCK, Walter, b. 1873.
The Bolinder book.
Stockholm: Bolinders, 1930. Frontis., xiii, 180, [1] (ads.); illd.   B
    Includes canal and river craft.

POLLOCK, Walter, b. 1873.
Building small ships.
Tunbridge Wells: Executors of Walter Pollock, 1948. Frontis., [i], 363; illd. B
    Includes canal and river craft.

POLLOCK, Walter, b. 1873.
Designs of small oil-engined vessels.
London: Constable, 1927, xiii, 86, illd., 95 fldg. pl.   B
    Includes canal and river craft.

POLLOCK, Walter, b. 1873.
Hot bulb oil engines and suitable vessels.
London: Constable, 1919, xix, 429, [2] (pub. cat.), 10 fldg. pl; illd.   B
    Includes canal and river craft.

POLLOCK, Walter, b. 1873.
The Pollock's as engineers.
Tunbridge Wells: Pollock, 1939. [i], 228; fldg. tables.   B
    Includes canal and river craft.

POLLOCK, Walter, b. 1873.
Small vessels.
Tunbridge Wells: Pollock, 1946. Frontis, [iii], 267; illd.   B
    Includes canal and river craft.

QUAESTOR, *pseud. See* BYFORD-JONES, W.

ROBERTS, A. W.
Coasting bargemaster.
London: Arnold, 1949. 192; drgs., map endpprs.   B
    Thames sailing barges.

SMITH, Emma.
Maidens' trip.
London: Putnam, 1948. [iv], 208.   B
    2nd imp. 1949. Pprbk edn. by Pocket Bks. (London) 1950. Large-print edn. by Chivers (Bath)
    1977. Abridged edns.: Chatto & Windus (London) 1953, 3rd imp. 1961; Penguin (Harmonds-
    worth) 1964, or pprbk.
    Women boat crews on Grand Union in WW2.

George Smith of Coalville: a chapter in philanthropy.
London: Haughton, [1881?]. Frontis., portrait title, 56, [4] (pub. cat.).   B

SMITH, George, 1831-1895.
Canal adventures by moonlight.
London: Hodder & Stoughton, 1881. Frontis., vi, 250.   B
   Missionary work amongst narrow-boat families.

SMITH, George, 1831-1895.
Our canal, gipsy, van and other travelling children; and the steps taken, and being taken, to improve their condition. A lecture by George Smith . . .
[Rugby?]: privately printed, 1888. 33.   B

SMITH, George, 1831-1895.
Our canal population: the sad condition of the women and children, – with remedy. An appeal to my fellow country men and women.
London: Haughton, 1875, 111; illd.   B
  New edn. [1879] subtitled 'a cry from the boat cabins, with remedy'. Facs. of new edn. by EP pubg. (Wakefield) 1974.

STEVENS, John Lee.
Description of a new method of propelling steam vessels, canal boats, &c. . . .
London: Glyn, 1828. [ii], fldg. frontis., 17.   B

TRENT NAVIGATION, Committee of Proprietors of.
A set of tables for ascertaining the weight of the cargoes carried by boats navigating on the River Trent . . .
Nottingham: PTN, 1799-1800. 2 vols. [not continuously paginated].   L

WILLIAMSON, James.
The Clyde passenger steamer: its rise and progress during the nineteenth century.
Glasgow: MacLehose, 1904. Frontis., xv, 382; illd.   B

WOOLFITT, Susan.
Idle women.
London: Benn, 1957. Frontis., 223; boat plan, map endpprs.   B
   Women boat crews on Grand Union in WW2.

6: Cruising; cruising guides; packet boat trips.

AUBERTIN, C[arey] J[ohn], 1876-1932.
A caravan afloat.
London: Simpkin et al., [1916]. Frontis., 155, [15] pl.   B
  Facs. edn. by Shepperton Swan (Shepperton, Middx.) 1982.
     Canal trips by unpowered houseboat.

BARNES, Eleanor [Cecilia], [afterwards Lady Yarrow].
As the water flows: a record of adventures in a canoe on the rivers and trout streams of southern England.
London: Richards, 1920. 190; illd.   B
  2nd edn. by Ingleby (London) 1927.
     Romantic account of trips on Arun, Avon (Hants.), Mole, Rother etc.

BENNETT, A[rthur] S[crivener].
June of Rochester: topsail barge.
London: Arnold, 1939. Frontis, 256, [15] pl.; illd.   B
  2nd edn. 1949.
     Conversion and travels of Thames sailing barge.

BENNETT, [Enoch] Arnold, 1867-1931.
From the log of the Velsa.
New York: Century, 1914. 307; illd.   N,P
  U.K. edn. by Chatto & Windus (London) 1920. Facs. of 1st edn. by Books for Libraries Press
  (New York) 1975.
        Includes sailing on East Anglian rivers.

BICKERDYKE, John, *pseud.* [Charles Henry COOK, 1858-1933].
The best cruise on the Broads, with useful hints . . .
London: Bliss *et al.*, [1895]. Frontis., 142, [6] (ads.), [7] pl.   B

BLACK, Malcolm.
'Our canoe voyage': how we launched at Hereford and landed at Reading.
Manchester: for private circulation, 1876. [i], 371.   B
    Cover title 'Minnie and Undine. Our canoe voyage'.
        Via R. Wye, Thames & Severn Canal.

BLISS, William.
Canoeing: the art and practice of canoeing on English rivers, navigations and canals: with a descrip-
tion and tables of distances of the canoeable waterways of England and Wales.
London: Methuen, 1934. xii, 284, 8 (pub. cat.), fldg. map.   B
    2nd edn. 1947.

BLISS, William.
The heart of England by waterways: a canoeing chronicle by river and canal.
London: Witherby, 1933. frontis., 192, [2] pl., fldg. map.   B

BLISS, William.
Rapid rivers.
London: Witherby, 1935. Frontis., xii, 244, [7] pl.; maps.   B
        Canoeing. Includes Brecon & Abergavenny Canal.

BONTHRON, P[eter].
My holidays on inland waterways: 2000 miles cruising by motor boat and pleasure skiff on the
canals and rivers of Great Britain.
London: Murby, [1916]. Frontis., xvii, 186, II-XXXII pl., fldg. map.   B
    3rd edn. 1919.

BOUMPHREY, Geoffrey [Maxwell].
Down river: a canoe tour on the Severn and Thames.
London: Allen & Unwin, 1936. Frontis., 127, II-XV pl. (Allen & Unwin Pocket Crown Series No. 12). B
        Includes passage along the abandoned Thames & Severn Canal.

BRITISH CANOE UNION.
Handbook and guide to the waterways of G. Britain & Ireland.
London: BCU, 1936. 126.   B
    [4th] edn. 1966; 2nd imp. 1970.

BRITTAIN, Harry.
Notes on the Broads and rivers of Norfolk & Suffolk.
Norwich: Argus Office, 1887. [16] (ads.), frontis., [viii], fldg. map, 154, [11] pl., [16] (ads.);
maps, illd.   B

BRITTAIN, Harry.
Rambles in East Anglia: or, holiday excursions among the rivers and Broads.
London: Jarrold, [1890]. Frontis., 150, [8] (ads.); illd.   B
    3rd edn. 1897.
        Includes trips by yacht, steamer, wherry etc.

BURRAGE, Douglas.
An old waterway revived.
Birdham: Chichester Yacht Co., 1936. 24; illd., maps.   P
    Cover title 'Safe mooring'.
        Refers to Chichester Canal.

BYFORD-JONES, W[ilfred] ;
Vagabonding through the Midlands.
London: Cranton, 1935. 208; illd., map.   B
     31–64 describe trip in horse-drawn narrow-boat from Wolverhampton to Llangollen.

Canal and river; a canoe cruise from Leicestershire to Greenhithe, including a guide to the Thames below Oxford, by Red Rover.
Bedford: Hill, 1873. [vi] , 90.   P

CARR, John, 1772–1832.
The stranger in Ireland; or, a tour in the southern and western parts of that country, in the year 1805.
London: Phillips, 1806. Fldg. frontis., xv, 530, 16 pl. (some fldg.).   B
     434–5 describe packet boat trip Athy to Dublin.

CARRINGTON, Noel, and CAVENDISH, Patricia. (eds.).
Camping by water.
London: Davies, 1950. Frontis., 140; illd., maps.   B
     Contains guide to many English waterways.

Chester–Kendal canal trip, 1899.
[No title page]. 130.   P
     By canoe.

CHILDERS, J[ohn] W[albancke], (ed.).
Lord Orford's voyage round the Fens in 1774.
Doncaster: White, 1868. [i], ii, 107.   B

CLARKE, J. F. Mostyn.
Three weeks in Norfolk: being a portion of the 'Rover's' log.
London: Wyman, 1887. Frontis., [iii], 53, [8] pl.   P

CLOUGH, S.
Cruises and curses.
London: Selwyn & Blount, 1926. viii, 254; drgs.   B
     Cruising on Thames and Regent's Canal.

COLES, K[aines] Adlard.
Creeks and harbours of the Solent (including Spithead).
London: Arnold, 1933. 96; illd., charts. 8th edn. 1972.   B

[COLQUHOUN, Patrick MacChombaich de].
A companion to The Oarsman's Guide. By the honorary secretary of the 'Leander Club'.
Lambeth: Searle, 1857. [iv], 32.   B

A companion to The Oarsman's Guide.
See COLQUHOUN, Patrick MacChombaich de.

COOKE, Francis B[ernard].
Coastwise cruising from Erith to Lowestoft.
London: Arnold, 1929. xi, 212, 16 (pub. cat.). drgs, maps.   B
     Includes R. Alde, Blackwater, Crouch, Deben, Orwell and Stour.

COOKE, Francis B[ernard].
London to Lowestoft: a cruising guide to the East Coast.
London: Imray et al., [c. 1906]. [vii], 172, [4] (ads.); illd.   B
     Includes R. Alde, Blackwater, Colne, Crouch and Deben.

COTES, V. Cecil.
Two girls on a barge.
London: Chatto & Windus, 1891. 2 (pub. cat.), viii, 177, 32 (pub. cat.); illd. B
   2nd edn. 1894.
     Trip by narrow-boat, London to Coventry.

CROSSLEY, Sydney.
Pleasure and leisure boating: a practical handbook.
London: Innes, 1899. viii, 256, viii (pub. cat.).   B

Cruises on some of the western lochs of Scotland, by five members of the Mersey Canoe Club.
Liverpool: Dobb (printer), [c. 1883]. Frontis., 46, [8] pl.   P

CUBBON, T. W.
'Only a little cockboat': roughing it from Dee to Severn and Avon and canals between.
London: Roberts, 1928. Frontis., 192, [17] pl.; map on rear endpprs.   B
    Later issued with label of Houghton Pubg. Co. (London) on title page.
        By motor launch.

CUBBON, T. W.
The wizard Dee: a June voyage from Bala to the sea.
London: Witherby, 1934. Frontis., 176, [4] pl.; map.   B
        By rowing boat and motor launch.

DASHWOOD, J. B[acon].
The Thames to the Solent by canal and sea, or the log of the Una boat 'Caprice'.
London: Longmans et al., 1868. Frontis., ix, fldg. map, 91, [6] pl.   B
    Facs. edn. by Shepperton Swan (Shepperton, Middx.) 1980.

DAVIES, G[eorge] Christopher, 1849–1922.
The handbook to the rivers & Broads of Norfolk & Suffolk.
London: Jarrold, 1882. [ii] (ads.), 108, [vi] (ads.); illd., fldg. map in pkt.   B
    50th edn. 1929 entitled 'Rivers and Broads . . .'.

DAVIES, G[eorge] Christopher, 1849–1922.
Norfolk Broads and rivers, or the water-ways, lagoons, and decoys of East Anglia.
Edinburgh: Blackwood, 1883. Frontis., ix, 290, [11] pl.   B
    2nd edn. 1884.

DAVIES, G[eorge] Christopher, 1849–1922.
The Swan and her crew, or the adventure of three young naturalists and sportsmen on the Broads
and rivers of Norfolk.
London: Warne, [c. 1876]. xx, 267, 32 (pub. cat.); illd.   B
    5th edn. [c. 1890], later edns. 1924, 1932.
        More natural history than boating.

DE SALIS, Rodolph Fane.
Thirty days on English canals, with some remarks on canal development.
Oxford: privately printed, 1894. 15.   P
        Cousin of Henry Rodolph De Salis.

DODD, Anna Bowman, 1855–1929.
On the Broads.
London: MacMillan, 1896. xii, 331; illd.   B
        A cruise.

[DOUGHTY, Henry Montagu].
Summer in Broadland: gipsying in East Anglian waters. By the author of 'Friesland Meres', &c.
London: Jarrold, 1889. Frontis, 136, [10] (pub. cat.); illd.   B
    6th edn. [1897], not anonymous.

Down the river from Oxford to Nuneham & Abingdon.
Oxford: Salter, 1923. 46 (inc. ads.); illd., map.   B

DOWNIE, R[obert] Angus, b. 1905.
The heart of Scotland by waterway: canoe adventure by river and loch.
London: Witherby, 1934. Frontis., 167, [5] pl.   B

DUTT, W[illiam] A[lfred].
A guide to the Norfolk Broads. Being Part I of 'The Norfolk Broads'.
London; Methuen, 1923. Frontis., [vii], 218, 8 (pub. cat.), [11] pl., fldg. map.   B
    See Section 2 for 'The Norfolk Broads'.

ELLIS, A[lexander] R[obert].
The book of canoeing: how to buy or make your canoe and where to take it.
Glasgow: Brown *et al.*, 1935. Frontis., xii, 197, [1] (ad.), [1] fldg. pl., [8] fldg. maps;
illd.   B
   3rd edn. 1946.

ELLIS, Alec [i.e. Alexander] R[obert].
Canoeing for beginners.
Glasgow: Brown *et al.*, 1936. 42, [6] (pub. cat.); illd.   (B),P
   2nd edn. 1946.

EMERSON, P[eter] H[enry], 1856-1936.
Idyls of the Norfolk Broads. Being twelve autogravure plates . . . with descriptive text . . .
London: Autotype Co., 1887. x, [12], [12] loose pl.   B

EMERSON, P[eter] H[enry], 1856-1936.
Marsh leaves.
London: Nutt, 1895. viii, 165, XVI pl. (inc. frontis.).   B
   Unilld. edn. by Perry (Stratford) 1898.
      Recollections of Norfolk Broads.

EMERSON, P[eter] H[enry], 1856-1936.
On English lagoons: being an account of the voyage of two amateur wherrymen on the Norfolk
and Suffolk rivers and Broads.
London: Nutt, 1893. Frontis., xii, 298, [14] pl., 2 (pub. ad.); illd.   B

EMERSON, P[eter] H[enry], 1856-1936.
Wild life on a tidal water: the adventures of a houseboat and her crew.
London: Sampson Low *et al.*, 1890. xiv, 145, XXX pl. (inc. frontis.), fldg. chart.   B
   Breydon water.

EMERSON, P[eter] H[enry], and GOODALL, T. F.
Life and landscape on the Norfolk Broads . . .
London: Sampson Low *et al.*, 1886. Frontis., 81, XXXIX pl.   B

EVANS, John, 1767-1827.
An excursion to Windsor, in July, 1810 . . . also a sail down the River Medway, July, 1811, from
Maidstone to . . . the Nore . . .
London: Sherwood *et al.*, 1817. Frontis., [ii], x, 558, [6] (ads.); illd.   B
   2nd edn., 1827, omits Medway trip.

EVERITT, Nicholas.
Broadland sport.
London: Everett, 1902. xxiv, 393, 22 (ads.); illd.   B
   Substantial portion on boating.

FARRAR, C[harles] F[rederick], b. 1860.
Ouse's silent tide.
Bedford: Sidney Press, 1921. Frontis., xi, 225, [57] pl., fldg. map   B
   2nd edn. 1922 but dated 1921 on title-page.
   Facs. edn. by SR Publishers (Wakefield) 1969.
      Canoeing on Gr. Ouse and tributaries.

FINCH, Robert J.
To the west of England by canal.
London: Dent, 1912. 64 pp., [12] pl.; illd., maps. (Educational Journey Series).   B
   Educational handbook resulting from four school trips on Kennet & Avon.

GEDGE, Paul.
Thames journey: a book for boat-campers and lovers of the river.
London: Harrap, 1949. Frontis., 144, [30] pl.; map endpprs.   B

GIBBINGS, Robert [John].
Sweet Thames run softly.
London: Dent, 1940. x, 230; illd.   B
   10th imp. 1946. RU edn. 1941, Guild Bks. edn. 1944.
      By rowing boat down to Windsor.

GRAY, Johnnie, *pseud*. [H. Speight].
A tourist's view of Ireland.
London: Simpkin *et al.*, *c.* 1885. 55.   B
      Trips on L. Corrib (44) and Killarney (53–4).

A guide to the Upper Thames from Richmond to Oxford, for boating men . . .
London: Gill, 1882. 103, [1] (pub. cat.); illd.   B
    For 2nd edn., *see* 'The Thames guide book'.

HALL, Mr. & Mrs. S[amuel] C[arter].
Ireland: its scenery, character, &c.
London: How & Parson, 1841-2-3. 3 vol. Frontis., xii, 435, [22] pl.; illd. viii, 468, [19] pl.; illd.
Frontis., 512, [20] pl.; illd.   B
   Further edns.: How, 1846; Hall, Virtue (London) *c.* 1865.
     Fly-boat trips on Grand Canal (II 190–1) and Royal Canal (III 275–6).

HAYWARD, John D[avey].
Canoeing with sail and paddle.
In BELL, Ernest, (ed.). Handbook of athletic sports.
Vol. VII: Canoeing and camping out.
London: Bell, 1893. [iii], 1–152; illd.   B
   Also issued as a separate volume.

HAYWARD, Richard.
Where the River Shannon flows.
London: Harrap, 1940. Frontis., 328, [30] pl., fldg. map.   B
   2nd edn. by Dundalgan Press (Dundalk) 1950.

HEAVISIDE, George.
Canoe cruise down the Leam, Avon, Severn and Wye.
London: Whitfield, 1871. vii, 3–102.   B

HOLDING, T[homas] H[iram], 1844–1930.
The cruise of the Ospray canoe and camp life in Scotland . . .
Newcastle-on-Tyne: Bell, 1878. Frontis., 61.   P

HOLDING, T[homas] H[iram], 1844–1930.
Watery wanderings 'mid western lochs: a practical canoe cruise.
London: Marlborough, 1886. 144, [1] (ad.); illd.   B

HORTON, C. (ed.).
Handbook of the Motor Boat Association.
London: MBA, 1930. viii, 295 (inc. ads.); illd.   P
    13–122 on inland waterways.

HORTON, C.
Motoring afloat.
London: Society of Motor Manufacturers & Traders, 1927. 175; illd.   B
    94–109 on British inland waterways.

HOUSLEY, S. J.
Comfort in small craft: a practical handbook of sailing and cookery.
London: Murray, 1911. [v], 128, [12] (ads.); drgs.   B
   Reissued by Blake's (London) 1925 and 1928, the latter edn. having a slightly altered title.
     Refers solely to Norfolk Broads and rivers.

INGLIS, Henry [David], 1795–1835.
A journey throughout Ireland, during . . . 1834.
London: Whitaker, 1834. 2 vol. Fldg. map frontis., xii, 349. Fldg. map frontis., viii, 348.   N
    5th edn., 1938.
        Boat trips on Blackwater (I 173–4), Lee (I 192–3) and Shannon (I 265–74, 318–33; II 9–14, 144–5).

INWARDS, James.
Cruise of the Ringleader.
London: Simpkin et al., 1870. Frontis., 126.   B
        Canoeing on Scottish inland and coastal waters.

IONIDES, Cyril, and ATKINS, J[ohn] B[lack].
A floating home.
London: Chatto & Windus, 1918. Frontis., xvi, 200, [15] pl.; plan.   B
        Thames sailing barge.

IRVING, John.
The yachtsman's pilot. Vol. 1. Rivers and creeks of the Thames estuary . . .
London: Saturday Review, 1927. [viii], 80, [2] (ads.).   B
    2nd (revd.) edn. by Watts (London) 1933.
        Includes R. Alde, Blackwater, Colne, Crouch, Deben, Medway, Orwell and Stour.

JENNINGS, Payne.
Sun pictures of the Norfolk Broads. One hundred photographs from nature of the rivers and Broads of Norfolk and Suffolk.
Ashtead, Surrey: Permanent Photo Printing Works, [c. 1891]. [iii], 50 leaves (printed one side only), [iii] (ads.).   P
        No text. See SUFFLING, Ernest Richard, for 2nd [1892] and 3rd [1897] edns.

JOHNSON, James, 1777–1845.
A tour in Ireland . . .
London: Highley, 1844. Frontis, xii, 372, [4] (pub. cat.).   B
    58–90 on journey on Grand Canal and Shannon.

JONES, H[enry] Lewis, and LOCKWOOD, C. B.
Swin, Swale and Swatchway: or, cruises down the Thames, the Medway and the Essex rivers.
London: Waterlow, 1892. Frontis., [iii], viii, map, 203, [19] pl.; illd.   B

KINGSTON, William H[enry] G[iles], 1814–1880.
A yacht voyage round England.
London: Religious Tract Soc., [1879]. 334, [2] (pub. cat.).   B
    New edn. [1890].
        251–258 on transit of Caledonian Canal.

KOHL, J[ohann] G[eorg], 1808–1878.
Ireland, Scotland and England.
London: Chapman & Hall, 1844. v, 248, (iii), 100, (i), 202.   B
        31–39: Shannon trip.

LAFFAN, Mrs. R. S. de Courcy, d. 1912.
The cruise of the 'Tomahawk'.
London: Eden et al., 1892. [ix], 132, [3] (pub. cat.); illd.   B
        Thames rowing holiday, Oxford to Cookham.

[LAING, Charles Coleman.]
A week on the Bure, Ant & Thurne.
[?]: privately printed, 1895. Frontis., [xv], 59, [3] pl., map.   P,N

LA TOCNAYE, Chevalier de.
Rambles through Ireland; by a French emigrant.
London: Robinson, 1799. 2 vol. ix, 240. [i], 251.   B
    Another translation by McCaw et al. (Belfast) and Hodges et al. (Dublin) c. 1917, under title 'A Frenchman's walk through Ireland'.
        21–23 canal trip Athy-Dublin.

LEDGER, Walter E[dwin].
The 'Blue Bird' among the Norfolk reeds, with some reflections on the water.
London: Roworth (printer), 1911; 25; mounted photo frontis.   B
    Limited edn. of 50.

LEYLAND, John, 1858[?]–1924.
The Thames illustrated: a picturesque journeying from Richmond to Oxford.
London: Newnes, [1897]. [iv], 284; illd.   B
    Issued in 12 weekly parts.

LLOYD, Montague, and LLOYD, Ann.
Through England's waterways.
London: Imray et al., 1948. Frontis., [vii], 117, [12] pl., loose map.   B

LOWNDES, George R.
Camping sketches.
London: Bentley, 1892. [vii], 247; illd.   B
        169–213 describe dinghy trip down Dorset Stour.

McCARTHY, R. H.
Canoeing.
London: Pitman, 1940. xii, 179, fldg. map; illd.   B
    2nd edn. 1951.

MACDONELL, Arthur A[nthony], 1854–1930.
Camping out.
In BELL, Ernest, (ed.). Handbook of athletic sports. Vol. VII: Canoeing and camping out.
London: Bell, 1893. [iii] , 1–153; illd., maps.   B
        Gives a lot of information on boat-camping on inland waterways.

MANNING, E[liza] F.
Delightful Thames.
London: Low et al., [c. 1896]. 29; illd.   B

MARSHALL, John.
Canoes and canoeing. With a guide to Scottish waters.
Stirling: Mackay, 1937. 37; illd.   B

MESSUM, S[tuart] V[ictor] S[eymour] C[raigie].
East Coast rivers. Charts and sailing directions for the Rivers Crouch, Roach, Blackwater, Colne,
Stour, Orwell, Deben, Ore and Alde . . .
London: Potter, 1903. Frontis., 42, 8 (pub. cat.), 11 pl. (some fldg.).   B

MORLEY, F[rank] V[igor].
River Thames.
London: Methuen, 1926. Frontis., xi, 255, [4] (pub. cat.), [15] pl., [9] maps.   B
        Cricklade to London by water.

MURRAY, Alison D.
Burrow's guide to the River Thames.
Cheltenham: Burrows (for Thames Boating Trades Assn.), [c. 1929]. 116, [8] pl., fldg. map; drgs.,
maps.   B

NEAL, Austin E.
Canals, cruises, and contentment.
London: Cranton, [1921]. Frontis., 189, [11] pl.   B

The oarsman's guide to the Thames . . .
See WOOD, Thomas Lett.

PASK, Arthur T[homas].
From lock to lock. A playful guide to the Thames, from Teddington to Oxford.
London: Judy, [c. 1882]. 126 (inc. ads.); illd.   B

PATTERSON, Arthur Henry.
The cruise of the 'Walrus' on the Broads: a Broadland voyage in the North Sea ketch-boat.
London: Jarrold, [1923]. Frontis., 175, [8] (ads.); illd.   B

PATTERSON, A[rthur Henry].
Man and Nature on the Broads.
London: Mitchell, 1895. 143; illd.   B

PATTERSON, Arthur Henry.
Through Broadland by sail and motor.
London: Blakes, 1930. Frontis., xiii, 141; drgs.   B

PATTERSON, Arthur H[enry].
Through Broadland in a Breydon punt. By 'John Knowlittle'.
Norwich: Vince, 1920. Frontis., xvi, 112, 20 (ads.), [2] pl.; drgs.   B

PENNELL, Joseph and PENNELL, Elizabeth Robins.
The stream of pleasure: a narrative of a journey on the Thames from Oxford to London.
London: Unwin, 1891. Frontis., 160, [2] (ads.); illd.   B

PROTHERO, F[rancis Thomas] E[gerton], and CLARK, W[illiam] A[bercrombie] (eds.).
Cruising Club manual. A new oarsman's guide to the rivers and canals of Great Britain and Ireland.
London: Philip, 1896. vi, 302, [2] (pub. cat.), fldg. map in rear pkt.   B

RAVEN-HART, R[owland James Milleville].
Canoeing in Ireland.
London: Canoe & Small Boat, [1938]. 40; illd., map.   P

[RICHARDS, Thomas Bryan].
A voyage up the Thames.
London: Roberts, 1738. 100.   B
    Also attrib. to Weddell.

A river holiday.
See SIMS, W.

ROBERTSON, E[ileen] Arnot.
Thames portrait.
London: Nicholson & Watson, 1937. xi, 174, [50] pl.   B
    Boat trip down Thames.

ROBINSON, Jack.
Broadland yachting: a practical handbook of sailing . . .
Oulton Broad: by the author, 1920. 98 (inc. ads.); illd., maps.   B
    3rd edn. 1922. [4th?] edn. 1928.

ROBINSON, Jack.
Motor cruising on the Broads . . . a practical handbook . . .
Oulton Broad: by the author, 1934. 110; illd., map.   B

ROBINSON, Leo A.
Yachting on the Broads.
Lowestoft: Robinson, 1926. [vii], 103 (inc. ads.), fldg. map; illd.   P
    At least 11 subsequent edns. 1932-1953.

ROLT, L[ionel] T[homas] C[aswall], 1910-1974.
Green and silver.
London: Allen & Unwin, 1949. 275, 24 pl., [3] maps.   B
    2nd (revd.) imp., 1968.
        Cruising on Irish waterways.

ROLT, L[ionel] T[homas] C[aswall], 1910-1974.
Narrow boat.
London: Eyre & Spottiswood, 1944. 212; drgs.   B
    2nd imp. 1945. RU edn. 1946. Revd. edn. 1948. Right Bk. Club edn. 1949.
    Pprbk edn. 1978; 2nd imp. 1980.
        Cruises on English waterways.

Royal Thames guide containing thirty-five diagrams from lock to lock . . .
London: Simpkin *et al.*, 1899. Frontis. map, xii, 210; illd., maps.   B
    5th edn. 1906.

SALTER, John H[enry].
A guide to the River Thames . . . with particulars of the Rivers Avon, Severn and Wye, and the
principal canals in connection.
London: Field & Tuer, [*c.* 1883]. [2] fldg. maps, [iv], xxi (ads.), iii–vii, 123, [1] (pub. cat.); illd.   B
    57th edn. *c.* 1960.

SCOTT-JAMES, R[olfe] A[rnold], b. 1878.
An Englishman in Ireland: impressions of a journey in a canoe by river, lough and canal.
London: Dent, 1910. Frontis., viii, 264, [9] pl., map.   B

[SIMS, W.].
A river holiday. [Signed W.S.].
London: Unwin, [*c.* 1883]. 23; illd.   B
    In verse.

SMITH, Cyril Herbert.
Through the Kennet and Avon Canal by motor boat.
London: Roberts, 1929. Frontis., [xiv], 77, map.   B

SNOXELL, [Frank] Harvey.
Motor boating.
London: Pitman, 1932. [ii] (ads.), vii, 143, [2] (pub. cat.), fldg. map; illd.   B

SPEED, H[arry] Fiennes, d. 1925.
Cruises in small yachts and big canoes; or notes from the log of the 'Watersnake['], in Holland and
on the South Coast, the logs of the 'Water Rat' and 'Viper', on the Thames and South Coast . . .
London: Norie & Wilson, 1883. Frontis. chart, viii, 288, [9] charts (2 fldg.); illd.   B
    For 2nd edn. *see* next entry.
        74–101 sailing canoe trip London to Henley and back.

SPEED, Harry Fiennes, and SPEED, Maude.
Cruises in small yachts, by Harry Fiennes Speed: and a continuation, entitled More Cruises, by
Maude Speed.
London: Imray *et al.*, 1926 (2nd edn.). Frontis., [xiv], 355, [14] pl., [4] charts (1 fldg.); illd.   B
    For 1st edn. *see* previous entry.
        'More Cruises' includes R. Arrow, Avon and Severn.

SQUIRE, John C[ollings].
Water-music, or a fortnight of Bliss.
London: Heinemann, 1939. [vii], 280.   B
    Reprinted (with another work) in 'Solo and Duet' by Reprint Soc. (London) 1943.
    Describes canoe trip with William Bliss (q.v.) on Midlands waterways.

SUFFLING, E[rnest] R[ichard].
How to organize a cruise on the Broads.
London: Jarrold, [*c.* 1891]. Fldg. map frontis., 140, [4] (ads.); illd.   P,(B)
    3rd edn. 1899.

SUFFLING, Ernest R[ichard].
The land of the Broads. A practical guide for yachtsmen, anglers . . .
London: Gill, 1885. Fldg. map, v, 80, [6] (pub. cat.).   B
    2nd (illd.) edn., 1887. Illd. edn. by Perry (Stratford) 1892. 7th edn. by Perry (Stratford) 1895.
    Later edns. have altered sub-title.

SUFFLING, Ernest R[ichard].
Sun pictures of the Norfolk Broads . . . by Payne Jennings.
Ashtead, Surrey: Jennings, [1892]. Frontis., [vi], 2–50, [49] pl., [2] (ads.).   B
    '2nd edn.' on title page. *See* JENNINGS, Payne, for 1st edn. 3rd edn. [1879].

Summer in Broadland. *See* DOUGHTY, Henry Montagu.

TANSEY, George.
Adventure by canal.
Stone, Staffs.: Victoria Printing Works (printer), [c. 1950]. 28 (inc. ads.); illd., map.   P
   Reprinted from articles in the *Daily Despatch*, Easter 1950.
      Hire cruiser trip on Trent & Mersey, Shropshire Union, and Staffs. & Worcs. Canals.

TAYLOR, John, 1580-1653.
John Taylor's last voyage, and adventure, performed from the twentieth of July last 1641 to the
tenth of September following. In which time he past, with a scullers boate from the citie of London,
to the cities and townes of Oxford, Gloucester, Shrewesbury, Bristoll, Bathe, Monmouth, and
Hereford . . .
London: for the author, 1641. [v], 26.   B

The Thames guide book, from Lechlade to Richmond.
For boating men . . .
London: Gill, 1882 (2nd edn.). viii, 144, [24] (ads.), 12 (pub. cat.), [4] (ads.); illd.   B
   Revised edn. of 'A guide to the Upper Thames' [q.v.] ;

THURSTON, E[rnest] Temple. b. 1879.
The 'Flower of Gloster'.
London: Williams & Norgate, 1911. Frontis., xi, 244, [5] pl., 16 (pub. cat.); illd.   B
   2 issues of 1st edn. — small 4to and large post 8vo. Unilld. edn. by Chapman & Hall (London)
   1918.
   Facs. edn. of 1st by David & Charles (Newton Abbot) 1968 with new plates; 2nd imp. 1972.

A trip through the Caledonian Canal and tour in the Highlands. By 'Bumps'.
London: for private circulation, 1861. Frontis., 290, [20] pl.   B
   Author probably Mr. Stockman, principal assistant to George Robert Stephenson, whose yacht
   made this trip.

TUNBRIDGE, Ernest.
High jinks: a yarn of the Norfolk Broads.
[Gt.] Yarmouth: [?], [1896?]. 136.   N

UNDERHILL, Arthur, 1850-1939.
A short history of the first half-century of the Royal Cruising Club.
London: for the RCC, 1930. [63], [6] pl.   (B),P

A voyage up the Thames. *See* RICHARDS, Thomas Bryan.

WAKEMAN, W[illiam] F[rederick], 1822-1900.
Three days on the Shannon; from Limerick to Lough Key.
Dublin: Hodges & Smith, 1852. viii, 55, 52 (pub. cat.); illd.   B
   Guide for steamer passengers.

The waterway to London, as explored in the 'Wanderer' and 'Ranger', with sail, paddle, and oar,
in a voyage on the Mersey, Perry, Severn, and Thames, and several canals.
London: Simpkin *et al.*, 1869. Frontis., 96, [12] pl.; maps.   B

A week in a wherry on the Norfolk Broads, by 'Blue Peter';
London: Leadenhall Press, [c. 1890]. Frontis. 140, 14 (pub. cat.); illd.   B

A week on the Bure, Ant & Thurne.
See LAING, Charles Coleman.

WEETON, [Ellen], b. 1776.
Journal of a governess 1807-1811. Edited by Edward Hall.
London: Oxford Univ. Press, 1936. [i], frontis, xxii, 351, [7] pl.   B
   Facs. edn. by David & Charles (Newton Abbot) 1969.
      162-6 describe packet boat trip on Leeds & Liverpool Canal.

WESTALL, George.
Inland cruising on the rivers and canals of England and Wales . . .
London: Westall, 1908. Map frontis., 264, xxiii (index), [24] pl., fldg. map.   B

WHITE, Walter, 1811–1893.
Eastern England, from the Thames to the Humber.
London: Chapman & Hall, 1865. 2 vols. [xiv], 304, [2] fldg. maps. [xiv], 315, [2] fldg. maps.   B
    I: 77–177 describe a Broads cruise.   II: 179–190 a cruise on R. Waveney.

WILSON, W[illiam] Eric.
The pilots guide to the Thames estuary and the Norfolk Broads for yachtsmen.
London: Imray *et al.*, 1934. xv (ads.), xi, 411, [19] pl., 18 charts (some fldg.); illd.   B
    2nd edn. [1950]. 3rd edn. (Thames only) 1960.
        Includes many local rivers and creeks, e.g., Ant, Colne, Lea, Medway, etc.

[WOOD, Thomas Lett].
The oarsman's guide to the Thames . . . By a member of the Leander Club.
Lambeth: Searle, [*c.* 1856]. [i], 68.   B
    2nd edn. (1857) includes other rivers.

7: *Fiction*

(a) Adult

AIKIN, J[ohn], & AIKIN, A[nna] L[aetitia] (afterwards BARBAULD).
Miscellaneous pieces, in prose.
London: Johnson, 1773. [iii], 219.   B
    3rd edn. 1792.
        79-87 'The canal and the brook. A reverie'.

ASHBY-STERRY, J[oseph], d. 1917.
The river rhymer.
London: Ham-Smith, 1913. xi, 244, [4] (pub. cat.).   B
        Poems on the Thames.

ASHBY-STERRY, J[oseph], d. 1917.
A tale of the Thames.
London: Bliss *et al.*, 1896. [vii], frontis., 259, [4] (pub. cat.), [14] pl.   B

BATES, L[eonard] M[aurice].
Tideway tactics.
London: Muller, 1947. 166; illd.   B
        Stories of Thames lighter tug.

BLACK, William, 1841–1898.
The strange adventures of a house-boat.
London: Sampson Low *et al.*, 1888. 3 vols. [iv], 266. [iv], 269. [iv], 256, 32 (pub. cat.).   B
    First pub. in *Illustrated London News* as serial January–June 1888. 6th edn. 1890. New revd.
    edn., 1 vol., 1893.
        By horse-drawn boat from Kingston to Newbury via Birmingham and Bristol.

ELLIOTT, W. Gerald.
Treasure on the Broads.
London: Jenkins, 1933. 312, [8] (pub. cat.).   B

HERBERT, A[lan] P[atrick], 1890-1971.
The water gipsies.
London: Methuen, 1930. [vii], 392, 8 (pub. cat.).   B
    Numerous imps. and edns.
        Set on Thames and Grand Union Canal.

HOWARD, Keble, *pseud.* [John Keble BELL], 1875–1928.
The fast gentleman: a tale of the Norfolk Broads.
London: Fisher Unwin, 1928. 250, [3] (pub. ad.).   B

JEROME, Jerome K[lapka], 1859-1927.
Three men in a boat.
Bristol: Arrowsmith, 1889. [v], 315, [3] (pub. cat.); illd.   P,B
   Over 100 edns. B copy defective.

KNOX, Ronald A[rbuthnott], 1888-1957.
The footsteps at the lock.
London: Methuen, 1928. viii, 248, 8 (pub. cat.); map endpprs.   B
   5th edn. 1936. Large print edn. by Lythway Press (Bath) 1978.
     Set on Thames.

LANDER, Harry [Longley].
Lucky bargee.
London: Pearson, 1898. vii, 286, 15 (pub. cat.).   B
     Partly set on Thames sailing barge.

LEVER, Charles [James], 1806-1872.
Jack Hinton, the guardsman.
Dublin: Curry, 1843. [i], frontis., xi, 396, [26] pl.; illd.   B
   Pub. in parts as Vol. I of 'Our Mess'. Several later edns. as a separate work.
     129-135 describe packet-boat trip on Grand Canal of Ireland.

MILLER, Drew.
Seen from a windmill: a Norfolk Broads revue.
London: Heath Cranton, 1935. 236; drgs.   B

MORRIS, William, 1834-1896.
News from nowhere . . . some chapters from a Utopian romance.
London: Reeves & Turner, 1891. [ii], 238.   B
   First pub. by Roberts (Boston, Mass.) 1890.
   Numerous subsequent edns.
     158-225 describe rowing trip up Thames.

PEACOCK, Thomas Love, 1785-1866.
Crochet Castle.
London: Hookham, 1831. [v], 300.   B
   Many edns.
     Includes pleasure boat trip Thames-Pontcysyllte.

PEMBERTON, Murray.
Barge ahoy! A tale of the Thames in wartime.
London: Meridian, [1946]. 128; illd.   B

READE, Amos.
Life in the cut.
London: Swan Sonnenschein, 1889. viii, 343.   P,(B)

REMENHAM, John, pseud. [John Alexander VLASTO].
The canal mystery.
London: Skeffington, 1928. 256.   B
     Opens with discovery of severed head in Regent's Canal.

ROLT, L[ionel] T[homas] C[aswall], 1910-1974.
Sleep no more: twelve stories of the supernatural.
London: Constable, 1948. vii, 162.   B
   2nd edn. by Branch Line (Hassocks, Sussex). 1974 with changed subtitle.
     Two inland waterway stories.

SNOW, C[harles] P[ercy], b. 1905.
Death under sail.
London: Heinemann, 1931. [viii], 334; [2] plans.   B
   Revd. edn. 1959. Another edn. by Penguin (Harmondsworth) 1963.
     Set on Broads.

SUFFLING, Ernest R[ichard].
The innocents on the Broads.
London: Jarrold, 1901. Frontis., 329, [7] (pub. cat.).   B
    Collection of stories, some on cruising.

TROLLOPE, Anthony, 1815–1882.
The Kellys and the O'Kellys . . .
London: Colburn, 1848. 3 vols. [ii], 298. [ii], 298. [ii], 285.   B
   Several edns.
     I, 159–162 describe packet boat trip Dublin–Ballinasloe.

TROLLOPE, Anthony, 1815–1882.
The three clerks.
London: Bentley, 1858. 3 vols. [iv], 340. iv, 322. iv, 334.   B
   Several edns.
     Parodies a lowly branch of the Civil Service as 'The Office of the Commissioners of Internal
     Navigation'.

ZORN, Fritz.
Bunce, the Bobby, and the Broads: a holiday yarn.
London: Jarrold, 1900 (2nd edn.). 262, [2] (pub. ads.); illd.   P
   3rd edn. 1900.

(b) Inspired by the work of George Smith

COLBECK, Alfred.
Dick of the 'Paradise'.
London: Sunday School Union, [c. 1893]. 96; illd.   B

GRAY, Annie.
The old lock farm. A story of canal life.
London: Sunday School Union, [c. 1888]. 192, 16 (pub. cat.); illd.   B
     177–192 form Appendix by George Smith on 'Our canal population'. Another issue lacks this
     Appendix, and the pub. cat. Also issued bound with 'Ailie Stuart'.

LESLIE, Emma.
Tom the boater: a tale of English canal life.
London: Religious Tract Soc., [1882]. 160, 16 (pub. cat.); illd.   (B),P
   (*see* illustration on facing page).

LESLIE, Emma.
The water waifs: a story of canal barge life.
London: Partridge, [c. 1882]. 103; illd.   B
   Serialised in *The Childrens' Friend*, 1882.

MEADE, L[illie] T[homas], 1854–1914.
Water gipsies; or, the adventures of Tag, Rag, and Bobtail.
London: Shaw, [1883] (new edn.). Frontis., 224, [5] pl., [16] (pub. cat.).   B
   1st pub. in *Sunday Magazine* Christmas Number 1878.

PEARSE, Mark Guy, 1842–1930.
Rob Rat: a story of barge life.
London: Woolmer, [c. 1885] (17th thousand). 89, [7] (pub. cat.); illd.   P
   Several edns.: 26th thousand [c. 1888], 31st thousand, n.d. Author's name not on title-page,
   only on cover of some edns.

SARGENT, G[eorge] H[ewlett].
Ned, the barge boy.
London: Religious Tract Soc., [c. 1892]. Col. frontis., 80, 16 (pub. cat.).   B

LEGGING THROUGH A TUNNEL.   *Page* 35·

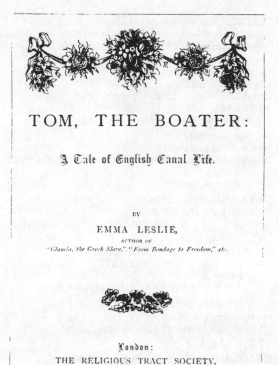

TOM, THE BOATER:

A Tale of English Canal Life.

BY

EMMA LESLIE,

AUTHOR OF

"*Glaucia, the Greek Slave,*" "*From Bondage to Freedom,*" *etc.*

London:

THE RELIGIOUS TRACT SOCIETY,

56, PATERNOSTER ROW; 65, ST. PAUL'S CHURCHYARD;
AND 164, PICCADILLY.

Figure 7.3. It seems unlikely that this artist had ever seen either a canal tunnel or a narrow boat, but these moral tales have an important place in the social history of canals. (*See* LESLIE, Emma).

STEAD, R[ichard].
Grit will tell.
London: Blackie, 1903. Frontis., 238, [3] pl., 32 (pub. cat.).   B

WILBRAHAM, Frances M.
Hal, the barge boy. A sketch from life.
London: Soc. for Promoting Christian Knowledge, [1883] . 73, 6 (pub. cat.); illd.   P,(B)
    In verse.

(c) Childrens'

Between the locks . . . *see* HOARE, E. N.

FRITH, Henry, b. 1840.
Aboard the 'Atalanta': the story of a truant.
London: Blackie, 1888 [but 1887]. [ii] (pub. cat.), frontis., 192, 32 (pub. cat.), [2] pl.   B
    40–81 on Regent's Canal.

HAMILTON, Frederick Spencer, b. 1856.
The holiday adventures of Mr. P. J. Davenant.
London: Nash, 1915. Frontis., vii, 160.   N
    Revd. edn. by Newnes (London) 1916.
        Includes plot to blow up royal train during its passage over Hardham tunnel (Arun Navn.).

[HOARE, Edward Newenham, b. 1842].
Between the locks; or, the adventures of a water-party.
London: SPCK, [*c.* 1877]. 127, 4 (pub. cat.); frontis.   B
    Moral tale based on pleasure trip up Medway.

HOGG, Garry [Lester].
Explorers afloat.
London: Nelson, 1940. Frontis., 314, 5 (pub. cat.); illd., map endpprs.   B
    2nd imp. 1952.
        Braunston–Brentford–Oxford.

HOGG, Garry [Lester].
House-boat holiday.
London: Nelson, 1944. Frontis., vii, 264 (inc. pub. ad.); illd., plan endppr.   B
    2nd imp. 1945.
        Set on Thames.

LEYLAND, Eric Arthur.
Discovery on the Thames.
London: Univ. of London Press, 1950. Frontis., 224; illd.   B

MITCHELL, Isla.
The beginning was a Dutchman.
London: Faber, 1944. 192; illd.   B
    3rd imp. 1945.
        Set on Thames and Grand Union Canal.

O'FARRELL, Brian.
Mystery on the river.
Oxford: Blackwell, 1936. Frontis., 256; illd. (Tales of Action series).   B
    2nd edn. 1943.

RANSOME, Arthur [Michell], 1884–1967.
Coot Club.
London: Cape, 1934. 352; illd., maps, map endpprs.   B
    Many edns.
        Set on Broads.

ROBERTSON, F. M.
The boatman's daughter.
London: Religious Tract Soc., [1896]. Frontis., 64, 16 (pub. cat.). (Little Dot Series).   P,(B)
    3rd edn. 1919.

SAVILLE, Malcolm.
The riddle of the painted box.
London: Carrington, 1947. 164; drgs.   B
    Several edns.
        Set on Grand Union Canal.

SAVILLE, Malcolm.
Two fair plaits.
London: Lutterworth Press, 1948. Frontis., 192; drgs., map endpprs.   B
    2nd edn. by May Fair Books (London) 1966.
        Set on Regent's Canal.

SEVERN, David, *pseud.* [David Storr UNWIN].
The cruise of the 'Maiden Castle'.
London: Bodley Head, 1948. 232.   B

WESTERMAN, Percy F[rancis], 1876–1959.
A mystery of the Broads.
London: Blackie, [*c.* 1930]. Frontis., 223, [3] pl.   B
    Reissued *c.* 1951. New edn. *c.* 1957.

WOOLFITT, Susan.
Escape to adventure.
London: Benn, 1948. 246; illd.   B
        Set on Grand Union Canal.

APPENDIX

A selection of anonymous tracts and pamphlets published before 1840

An address to the good sense of the country, on the projected Stamford Junction Navigation. By an inhabitant of Stamford.
Stamford: Drakard (printer), [c. 1810] 2nd edn. 52.   B

An address to the public, on the new intended canal from Stourbridge to Worcester, with the case of the Staffordshire and Worcestershire Canal Company.
[?]: [?], 1786. Fldg. map, 24.   B

An answer to a book intituled 'An inquiry into facts and observations thereon, humbly submitted to the candid examiner into the principles of a Bill . . . for the preservation of the Great Level of the Fens, and the Navigation through the same, by a tax on the lands and a toll on the navigation . . .'
London: for Cadell, 1778. [i], 130.   B

An answer to Mr. Fords booke, entituled A designe for bringing a navigable river, from Rickmansworth in Hartfordshire to St Giles in the Fields.
London: [?], 1641. [11].   B

An answer to the Worcester letter, dated December, 1785.
[?]: [?] [c. 1785]. 7.   B
        Supports Staffs. & Worcs. Canal Co. against proposed Worcester & Birmingham Canal.

An authentic description of the Kennet & Avon Canal. To which are added, observations upon the present state of the inland navigation of the south-western counties of England: and of the counties of Monmouth, Glamorgan, and Brecon, in south Wales.
London: Richardson, 1811. [iii], 30.   B
    Facs. edn. by K & AC Trust (London) 1969.

Avona; or a transient view of the benefit of making rivers of this kingdom navigable. Occasioned by observing the scituation of the city of Salisbury, upon the Avon, and the consequence of opening that river to that city . . . by R.S.
London: Courtney, 1675. [i], 33.   B

Brief remarks on the proposed Regent's Canal. By an observer.
London: Hatchard, 1812. 22.   B

A case against the junction of the English and Bristol Channels by a ship canal . . . by a subscriber of one share.
London: for the author, 1825. xi, 127.   B

The case of the barge-masters and others, navigating on the Rivers of Isis and Thames, from Oxford to London; shewing what hardships they labour under, by the exorbitant sums they pay for passing thro' the several locks, wears, bucks, gates, and for the use of boats belonging to the same, and going over towing-paths on the banks of the said rivers.
[?]; [?], [c. 1704]. 3.   B

The case of the citizens of Chester, in answer to several petitions from Leverpool, Parkgate, and the Cheesemongers; and also to printed reasons against the Act to recover and preserve the navigation of the River Dee.
[?]: [?], [c. 1734]. Single sheet.   B

The case of the Corporation of the Great Level of the Fenns; relating to a Bill depending in Parliament, for the better preservation of the Navigation of the Port of Kings-Lynn; which Bill is for taking away the Sluce at Denver-Dam . . .
[?]: [?], [c. 1660]. Single sheet.   B

The case of the navigation of the River Wye [i.e. Wey] in the county of Surry.
[?]: [?], [c..1660]. Single sheet.   B

The case of the proprietors of the Birmingham Canal Navigation, relative to Charles Colmore, Esq.
Birmingham: BCN, 1771. 4.   B

The case of the Town and Port of King's-Lynn in Norfolk, as to their navigation.
[?]: [?], [c. 1660]. Single sheet.   B

Considerations on the idea of uniting the rivers Thames and Severn through Cirencester, with some observations on other intended canals.
London: [?], 1782. [i], 21.   B

Considerations on the probable commerce and revenue, that may arise on the proposed canal, between Newcastle and Maryport.
Newcastle: Whitfield, 1796. 22   N

Considerations on the proposed cut from the Medway to the Thames . . .
London: Longman, 1827. Fldg. map frontis., 31.   B

Considerations on the scheme for . . . a navigable canal from the River Trent at Wilden Ferry in the county of Derby, to the River Mersey.
[?]: [?], [c. 1766]. 8, fldg. pl.   B

Considerations upon the intended navigable communication between the Friths of Forth and Clyde. In a letter to the Lord Provost of Edinburgh . . .
Edinburgh: [?], 1767. 21.   B

Copies of three letters upon the proposed navigable communication between the Friths of Forth and Clyde. Published in the Edinburgh News-papers in April 1767.
[Edinburgh]: [?], [1767]. 16.   B

A cursory view of a proposed canal: from Kendal, to the Duke of Bridgewater's canal, leading to . . . Manchester . . .
[?]: [?], [c. 1773]. 56.   B
            N records copy with fldg map. Attrib. to P. P. Burdett and R. Beck.

A cursory view of the advantages of an intended canal from Chesterfield to Gainsborough.
[?]: [?], 1769. [i], 18.   B

English and Bristol Channels Ship Canal. Prospectus and Mr. Telford's preliminary report, dated 2d August, 1824.
London: Brooke (printer), 1824. 10, 8.   C

An essay on ways and means for improving the inland navigation and increasing the number of sailors in Great-Britain.
London: Roberts, 1741. 37.   N

Extracts from the navigation rolls of the Rivers Thames and Isis. With remarks, pointing out the proper methods of reducing the price of freight. By a Commissioner.
London: Bathurst, 1772. 27.   B

Facts and reasons tending to shew, that the proposed canal, from the Trent to the Mersey, ought not to terminate at Northwich and Burton; . . .
[?]: [?], [c.1766]. 16.   B

Hydrographia Hibernica: or, a view of the considerable rivers of Ireland. Being an abstract of a . . . method of making them navigable . . .
[Dublin?]: [?], 1710. 15.   B

A letter to a friend, containing observations on the comparative merits of canals and railways occasioned by the reports of the Committee, of the Liverpool and Manchester Railway. [Signed 'F.P.']
London: Longman, 1832. [i], 32.   B
   2nd edn. 1832 'with additions'.
      By Frederick Page.

Letter to James Caldwell, Esq., on canals and rail-roads. [From 'a proprietor of the Grand Trunk Canal'.]
London: Staunton, 1825, 2nd edn. 10.   C
   1st edn. not seen.

A letter to Mr. J. J. Fox . . . exposing the injustice of his observations on the projected Stamford Junction Navigation . . . By an inhabitant of Stamford.
Stamford: Drakard, 1810. 19.   B

A letter to the Commissioners of the South Forty-foot Drain, in Lincolnshire, on their endeavours to obstruct the proposed . . . Stamford Junction Navigation. By an inhabitant of Stamford.
Stamford: Drakard (printer), 1810. [i], 10.   B

A letter to the Rev. Joseph Monkhouse and the Rev. Doctor Maurice Johnson on . . . the Stamford Junction Navigation.
London: Galabin & Marchant, 1811. 32.   B

A letter to William Poulteney, Esq.; on the subject of the Forth and Clyde Navigation. From a proprietor.
Edinburgh: [?], 1768. 29.   B

News from the Fens, or, an answer to a pamphlet entituled, Navigation prejudiced by the Fen-drainers. (Published lately in defence of the petition of Lin . . .).
London: [?], 1654. [i], 27.   B

Observations as to the powers of the Magistrates and Town Council of the City of Perth, in regard to the navigation of the River Tay.
Perth: [?], 1833. 26.   C
     Attrib. to John Miller.

Observations on a pamphlet entitled Thoughts on the proposed navigation between the Clyde and Carron.
[?]: [?], [c. 1767]. 11.   B

Observations on the account of a plan for the better supplying the cities of Edinburgh and Glasgow with coal, by Henry Steuart . . . By an old coal-master.
Edinburgh: Hill, 1800. [i], 63.   P,N
     By James Dunlop.

Observations on the general comparative merits of inland communication by navigations or rail-roads, with particular reference to those projected or existing between Bath, Bristol, and London: in a letter to Charles Dundas, Esq. M.P., chairman of the Kennet and Avon Canal Company. [Signed by 'A proprietor of shares in the Kennet and Avon Canal'.]
London: Hatchard, 1825. [i], 62, fldg. pl.   B

Observations on the intended Stamford Junction Canal.
London: Barry (printer), 1811. [i], 8.   B

Observations on the intended Stamford Junction Navigation . . . By an inhabitant of Stamford.
London: Philanthropic Soc. (printer), 1811. [i], 30.   B

Observations on the intended Worcester Canal.
[?]: [?], [c. 1785?]. 27, fldg. map.   B

Observations on the proposed Grand Union Canal.
London: Brooke (printer), [c. 1800]. [2].   B

A prophecy of Merlin. An heroic poem. Concerning the wonderful success of a project now on foot to make the river from the Severn to Stroud in Glocestershire navigable.
London: Bew *et al.*, 1776. [i], ii, 23.   B

Prospectus of the advantages to be derived from the Crinan Canal.
[?]: [?], [*c.* 1792]. 15, fldg. pl.   B

Prospectus of the projected Stamford Junction Navigation.
[?]; [?], [*c.* 1810]. Fldg. map, 7.   B

Reasons for making navigable the rivers of Stower and Salwerp, and the rivulets and brooks running into the same, in the counties of Worcester and Stafford.
[?]: [?], [*c.* 1670?]. Single sheet.   B

Reasons for making the River Dunn in the West-riding of the County of York navigable, and the great advantages which will accrue to the nation in general by it.
[?]: [?], [*c.* 1726?]. Single sheet.   B

Reflections on the general utility of inland navigation . . . with observations on the intended canal from Birmingham to Worcester, and some strictures upon the opposition given to it by the proprietors of the Staffordshire Canal. [Signed 'Publicola'.]
London: Debrett, [1798?]. 17, [3] pl. (2 fldg.).   B

Remarks on an intended navigable canal, from the River Trent at Wilden Ferry, in the county of Derby, to the River Mersey, in the county of Chester.
[?]: [?], [*c.* 1766]. 19.   B

Remarks on the supplement to an account of a plan for the better supplying the city of Edinburgh with coal, and on the examination comprised therein, by Henry Steuart . . . By the author of Observations of an old coalmaster.
Edinburgh: Hill, 1801. [iii], 92.   P,N
     By James Dunlop.

Remarks on the tonnage rates & drawbacks of the Grand Junction Canal, with observations on the proposed London and Birmingham Canal, by Mercator.
London: Simpkin, Marshall, 1836. 28.   C

Seasonable considerations on a navigable canal intended to be cut from the River Trent, at Wilden Ferry, in the county of Derby, to the River Mersey, in the county of Chester.
[?]: [?], [*c.* 1766]. 43, [2] fldg. maps.   B

Short and new state of the case, respecting the expence and usefulness of the small canal now under the consideration of Parliament, and of the larger canal proposed by Mr. Smeaton, between the Friths of Forth and Clyde.
Edinburgh: [?], 1767. 10.   B

State of inland navigation in Ireland . . .
Dublin: Shea (printer), 1810. 39.   N

Suggestions . . . [to] . . . the Magistrates & Town Council of the city of Perth, in exercising their powers in regard to the navigation of the River Tay . . .
Perth: [?], 1833. 22.   C
     Attrib. to John Miller.

Thames Navigation. A printed Bill has lately been handed about, intituled, Some few of the many objections that occur to the Bill now depending in Parliament, for making a navigable cut from Sunning in Berks, to Monkey Island, and from thence to Isleworth in Middlesex. And as the same tends to mislead the publick . . . it seems necessary to make some reply thereto . . .
[London?]: [?], [*c.* 1771]. 3.   B

Thoughts on the intended navigable communication between the Friths of Forth and Clyde. In a letter to his Grace the Duke of Queensberry, from a citizen of Edinburgh.
Edinburgh: [?], 1768. 27.   B

Thoughts on the navigation between the Friths of Forth and Clyde, and a navigable cut through the peninsula of Cantyre . . . from a Citizen of Edinburgh.
London: [?], 1777. 16.   G

To the subscribers of the intended Edinburgh and Glasgow Union Canal, and to the inhabitants of the city of Edinburgh. [Signed 'A Citizen'].
Edinburgh: Oliver & Boyd (printer), 1814. 12.   C

To Mr. Alderman Crisp Brown, of Norwich [a letter *re* developing navigation to Norwich].
Yarmouth: Meggy (printer), 1827. Single sheet.   B

A view of the advantages of inland navigations: with a plan of a navigable canal . . . between the ports of Liverpool and Hull.
London: for Becket *et al.*, 1765. [iii], 40, 2 fldg. maps.   B
    2nd edn. 1766.
        Attrib. to Thomas Bentley.

Western Union Canal, from the Thames near Maidenhead, to the Grand Junction Canal near Cowley. Replies to the arguments of the committee of the Thames Navigation Commissioners.
[London?]: Smeeton (printer), [*c.* 1820]. Fldg. map frontis., 22.   P

INDEX

LIST OF SUBSCRIBERS

Abophobia
Robert Aikman
D. C. Allen
F. V. Appleby
Professor J. H. Appleton
Keith M. Atkins
Robin Atthill
Peter Austin
Philip M. Backhouse
R. C. & D. M. Bailey
Terence E. Balchin
Myfanwy Baldwin
Jon Balley
Trevor A. Barker
S. A. Barnes
W. V. Barnes
Doreen Barratt
Arthur Barron
Douglas Batchelor
Kevin Batchelor
Alastair Bates
Nicola Bates
John W. Baylis
Jeremy M. Beales
Bruce A. Bearne
N. B. Benbow
A. F. Bensted, BCNS
Mayrice I. Berrill
Brian W. F. Berrington
John Berry
Derek Bews
David Bick
Gordon Biddle
Dawn Bijl
G. M. Binnie
G. L. H. Bird
David H. Boakes, C Eng, MI, Mech E
S. T. & T. C. J. M. Boddington
Harold Bode
J. D. Bolter
Roger Bonsall
The Book House
Nicholas Bostock
Joseph Boughey
John T. G. Bowen
John H. Boyes
R. Bradley
Elsie L. Bradshaw
Geoffrey Branson
A. L. Briggs
John V. Brightley
The British Waterways Board
Alan Elyard Brown
Joyce Brown
J. T. Bruce & family

R. A. Buchanan
Dr. John Bull
K. J. Burrow
David Burrows
D. J. L. Butcher
B. E. G. Byne, K & ACT, RST
Canoe Camping Club
Anthony Capo-Bianco
Nick & Marjorie Carter
K. E. & B. Catford
D. L. Chapman
P. W. K. Chapman
R. Chester-Browne
Christ's Hospital
Civil Engineering Library, Imperial
    College of Science & Technology
Richard Clarke
J. A. Clegg
Fraser Clift
Roger Cline
Stan Clover
Mrs. I. M. Codrington
Peter Collins
Philip Cohen
William John Cotterell
Dr. Edwin Course
David M. Cousins
Sharon A. Cox
Derek J. Crane
'Staff' Cripps, SHCS
Paul Gerard Crossland
L. J. Dalby
Alan Dance
Philip Daniell
Hector Davie
C. N. Davies
John C. Davies
Tony Davies (Norwich Branch IWA)
Daystar Theatre
Lionel Deeley
G. H. Dencer
John Dodwell
Frederic Doerflinger
Pauline Dower
Clive F. Drew
Roger A. Dunbar
Frank Dunham
Herbert R. Dunkley
J. E. Dunn
Dr. J. W. Eaton
J. K. Ebblewhite, ATI
John Edwards
L. A. 'Teddy' Edwards, MISTC
P. H. Edwards
Julia Elton

John R. Emerson
Dr. Michael Essex-Lopresti
Graham Ettles
J. M. Evans
Michael John Evans
Roger Evans
Robert C. Everett
Keith Fairclough
Keith Falconer
Alan Faulkner
J. B. Fautley
Clive Fazakerley
Malcolm H. Fielding
Ann Fileman
J. C. & M. Fletcher
Edmund J. Fogden
Jeremy Frankel
W. M. Franklin
John Freeman
Eric Garland
Eily Gayford
Michael Gilkes
Gloucester Road Bookshop, London
Gillian Goddard
Michael Golds
Kenneth Goodwin
O. H. Grafton, OBE, ASCA
M. & M. Grant
Melville & Judy Gray
Charles Anthony Green
Judith A. Grice, ALI, British Water-
    ways Board
Bruce Griffiths
Pam Griffiths
William J. Grose
David & Elizabeth Grove
C. W. Groves
Duncan L. Grundy
Martin Grundy
Mr. & Mrs. A. C. F. Hadfield
Alice Mary Hadfield
Robert Edward Hadfield
Bernard & Janet Hales
John Hankinson
Sheilagh A. Hardy
Frank E. Harper
Helen Harris
Paul Harrison
D. J. Harvey
Frances Mary Haydon
Arthur C. Hazzard, BEM
John Heap
Ruth Heard
David Heathcote
Dr. F. J. Hebbert

Heldite Ltd.
Lawrence D. P. Hicks
A. J. Higgitt
David Higham Associates Ltd.
David Hilling
Mr. & Mrs. A. K. Hirst
John & Jeanne Hodgkinson
J. R. H. Hogwood
Michael J. Holland
Stanley A. Holland
Frank & Margery Holroyd
M. G. Hooper
A. R. Hopkins—*Jason's Trip*
Brian A. Horsley
L. A. House
Humphrey G. W. Household
Michael Edward Huckstepp
David G. Hughes
John Humble
George Hume
Robert Humm
Denys Hutchings
J. N. Hutchinson
Guthrie Hutton
Michael Hyde
Alan Jeffery
A. M. Jervis
Brian Johnson
Rob Jones
Keith F. V. Joyce
Margaret Kay
A. M. Kerry, PPFAS, FRSA
A. C. Kidd
Philip Knight
J. M. Knott
David P. Lees
Robert & Mary Legget
M. C. Lewis
Dr. M. J. Lewis
John Peter Leyland
Jean Lindsay
Linlithgow Union Canal Society
Colin Livett
S. C. Lloyd
Hilda Mary Lomas
Eric Lomax
Kenneth S. Lord
Dr. R. Lorenz
Alan E. Lucas
Capt. H. W. Lucas, MN
J. Hunter MacDonald
William Joseph MacDonald
Ross I. Mackie
Mrs. Barbara MacMorran
Emily A. & J. Hayward Madden
J. Kenneth Major
C. J. Malcolm
Nigel Manning
Christopher M. Marsh
Harry & Nina Marsland
Mr. & Mrs. B. Martin
Dudley Matthews
P. H. Mattocks, MBE
Grenville C. Messham
Mikron Theatre Company
Michael G. Miller
Mr. & Mrs. R. B. Miller
Brian Mills

E. A. Mitchell—*Lady Fiona*
P. & D. L. Monahan
Bev Wm. Morant
P. T. Llewelyn Morgan
D. J. H. Morris
David N. Morris, BSc (Hons)
J. T. A. Morris
Roderick I. Muttram
Peter Myall
The Narrow Boat Co. Ltd.
National Waterways Transport
   Association
Lawrie Nelson
Nigel L. New
John G. H. Nicholls
Michael Norris
Peter Northway
Colin M. Notman
R. O. Nott
Mr. & Mrs. R. Oakley
Philip H. Ogden
B. C. Olivey
Liz & David Osborn
Bruce Edward Osborne
George Ottley
Courteney Owen
Dr. & Mrs. David Owen
Michael & Valerie Oxley
Richard C. Packer
S. M. Page
James R. Paisley
Martin Palmer
John Parkes, MBIM
W. J. Dennis Parkhouse
Ken Parris
Kenneth G. Parrott
Helen Provost Russell Parry
Raymond John Parsons
I. R. Patterson
Sheila Peacock
John & Marion Pearse
W. Pearse Chope
P. R. Pedlingham
Pensax
L. H. Phillips
Maurice W. Philpot
Alfred W. Picknell
D. C. Pinnock
Neil Pitts
W. E. & M. A. Platts
Robert S. W. Pollard
Miss A. M. Poole
Robin Blaize Porter
Arthur Pryor
Richard Pryor
P. R. Purcell
J. R. C. Quibell
Farrand Radley
G. A. Rawlins
D. M. R. Rees
Willie Reid
D. Richards
Peter Ricketts, Neath & Tennant CPS
Riparian Films
Michael Robbins
David Roberts
J. C. Roberts
Barbara Robinson

Leslie D. Robinson
Sonia Rolt
John F. Roper
Richard F. G. Ross
John V. Rotsey-Smith
Neil C. Rumbol
Ronald & Jill Russell
R. J. Rust
F. A. & S. M. Rymell
Carol & George Ryon
David Salt
David J. Sanders
J. R. Saunders
Pat Saunders
Peter Saw
Linda Jane Scott
Colin Scrivener
John Leslie Scrivener
Brian & Marty Seymour
Crispin M. Shann
Elizabeth M. Shaw
Edwin A. Shearing
K. B. Sherwood
David Shorto
Jack Simmons
John Simpkins
Frances M. Simpson
Sleaford Navigation Society
Marilyn Slinn
Bernard L. Smallman
David J. Smith
M. P. Smith
Neil A. Smith
Dr. Norman Smith
Raymond Smith
Robert Smith
Mark Smyth
A. J. & M. S. Snow
Peter M. Somervail
N. B. Sorceress
Jill Southam
Philip Spencer
Alan J. Squires
Alan Stevens
Michael L. Stevens
David Stevenson
J. L. & D. M. Stevenson
Peter Stevenson
Alastair Stewart
Michael A. Stimpson
A. I. Stirling, C Eng
Michael Streat
Joan Stretton
Thomas Robert Stuart
Capt. Dr. Thomas F. Swiftwater
   Hahn, USN (Ret.)
Hirokazu Tada
Christopher Tattersall
G. P. Taylor
Peter A. Taylor
G. W. Teed
David Telford
David H. Tew
James E. Theodore
C. Rhodes Thomas
Hugh Thomas
David St John Thomas, David &
   Charles